D1482937

Six Sigma for Technical Processes

Six Sigma for Technical Processes

An Overview for R&D Executives, Technical Leaders, and Engineering Managers

Clyde M. Creveling

PRENTICE
HALL

An imprint of Pearson Education

Upper Saddle River, NJ • Boston • Indianapolis • San Francisco
New York • Toronto • Montreal • London • Munich • Paris • Madrid
Cape Town • Sydney • Tokyo • Singapore • Mexico City

The publisher offers excellent discounts on this book when ordered in quantity for bulk purchases or special sales, which may include electronic versions and/or custom covers and content particular to your business, training goals, marketing focus, and branding interests. For more information, please contact:

U.S. Corporate and Government Sales
(800) 382-3419
corpsales@pearsontechgroup.com

For sales outside the United States, please contact:

International Sales
international@pearsoned.com

This Book Is Safari Enabled

 The Safari® Enabled icon on the cover of your favorite technology book means the book is available through Safari Bookshelf. When you buy this book, you get free access to the online edition for 45 days. Safari Bookshelf is an electronic reference library that lets you easily search thousands of technical books, find code samples, download chapters, and access technical information whenever and wherever you need it.

To gain 45-day Safari Enabled access to this book:

- Go to http://www.prenhallprofessional.com/safarienabled
- Complete the brief registration form
- Enter the coupon code U2JI-RLGK-NGQC-MIKQ-GUB1

If you have difficulty registering on Safari Bookshelf or accessing the online edition, please e-mail customer-service@safaribooksonline.com.

Visit us on the Web: www.prenhallprofessional.com

Copyright © 2007 Pearson Education, Inc.

All rights reserved. Printed in the United States of America. This publication is protected by copyright, and permission must be obtained from the publisher prior to any prohibited reproduction, storage in a retrieval system, or transmission in any form or by any means, electronic, mechanical, photocopying, recording, or likewise. For information regarding permissions, write to:

Pearson Education, Inc.
Rights and Contracts Department
One Lake Street
Upper Saddle River, NJ 07458
Fax: (201) 236-3290

ISBN 0-13-238232-6
Text printed in the United States on recycled paper at R.R. Donnelley and Sons in Crawfordsville, Indiana.
First printing, November 2006

Library of Congress Cataloging-in-Publication Data

Creveling, Clyde M., 1956-

Six sigma for technical processes : an overview for R&D executives, technical leaders, and engineering managers / Clyde M. Creveling.

p. cm.

ISBN 0-13-238232-6 (hardback : alk. paper) 1. Six sigma (Quality control standard) 2. Total quality management. I. Title.

TS156.C746 2006

658.4'013--dc22

2006027554

PRENTICE HALL SIX SIGMA FOR INNOVATION AND GROWTH SERIES

Clyde (Skip) M. Creveling, Editorial Advisor,
Product Development Systems & Solutions Inc.

MARKETING PROCESSES

Six Sigma for Marketing Processes
Creveling, Hambleton, and McCarthy

TECHNICAL PROCESSES

Design for Six Sigma in Technology and Product Development
Creveling, Slutsky, and Antis

Tolerance Design
Creveling

Six Sigma for Technical Processes
Creveling

What Is Six Sigma for Technical Processes? (Digital Shortcut)
Creveling

PRENTICE HALL SIX SIGMA FOR INNOVATION AND GROWTH SERIES

What Six Sigma has been and is becoming has stimulated an exciting, new body of knowledge. The old form of Six Sigma is all about finding and fixing problems using the ubiquitous DMAIC process. Cost savings and defect reduction are its goals. Financial returns from DMAIC projects occur at the bottom line of the financial ledger. Many would agree that the DMADV process was the next logical step in the evolution of Six Sigma methodologies when we need to design a new business process. The five step models have served us well, but it is time to look into the future.

The new form of Six Sigma uses tools, methods and best practices which introduces an approach to efficiently produce growth results within your company's existing business processes. It is not focused on reactive problem-solving, but rather on prevention of problems during the work you do to innovate, refresh and invigorate your business. Its financial return is at the top-line of the financial ledger. Its goal—innovation leading to growth.

Many are asking what is the future of Six Sigma? Has it run its course? Our experience tells us that Six Sigma is alive, evolving and expanding to meet new market demands. This is why we are introducing a new series of books that seeks to communicate a newly emerging branch of Six Sigma that focuses on creativity and new business growth. The **Prentice Hall Six Sigma for Innovation and Growth Series** contains books in two general process arenas: Marketing processes and Technical processes. The books in this series will span strategic, tactical, and operational process arenas to transcend the ongoing activities within a business. Product Portfolio Renewal and R&D are strategic processes; Product and Service Commercialization are tactical processes and Post-Launch Product Line Management, Sales, Customer Service and Support as well as Technical Production, Technical Service, and Support Engineering are operational processes. These processes would benefit from the rigor and discipline that Six Sigma-enhanced work produces.

Expect great things from this new series of books if you are looking for ideas on how to improve innovation and growth on a sustainable basis. They will take you to the next level of learning and doing through Six Sigma enablement within your organization. Classic Six Sigma is serving us well on the cost-side and we see only good news on the horizon through the evolution of Six Sigma for Innovation and Growth!

Clyde (Skip) M. Creveling, Editorial Advisor,
Product Development Systems & Solutions Inc.

This is the fifth book I have written, and, as always, I could not get these things done without the support, care, and feeding of my wonderful wife, Kathy; my patient son, Neil; and the two other women in my life to whom I owe a great deal for keeping me sane, out of jail, and at the right place at the right time, Ms. Carol Biesemeyer and Ms. Laura Schoenl. All my love and thanks.

I would also like to dedicate this book to a select group of true technical process heroes I have worked with over the years who do the hard work every day in their day-to-day work processes to get great technologies and products defined, developed, and out the door with as little reactive problem solving as possible:

Mr. Ernest Lifferth—Cummins, Inc.
Mr. Jerry Metcalf and Dr. Phil Miller—3M Corp.
Mr. Mo Vaheb, Dr. Hamid Fallahi, and Mr. Hussein El Habhab—
  Carrier Corp.
Mr. Pete Romeo, Mr. Bill Erdmann, Mr. Andy Jaunzemis, and Mr. Tom
  Borgoyn—Beckton Dickinson & Co.
Mr. Ken McIntyre and Mr. Bob Moroz—Merck & Co.
Dr. Dan Tegel—Seagate
Mr. Tom Judd—Motorola
Mr. Mike Kozack—Boeing Corp.
Mr. Jim Wasiloff—General Dynamics
Dr. Joe Sullivan—Logitech Corp.
Mr. Pat Hale—MIT & INCOSE
Dr. Steve Sanders—StorageTek
Mr. Jeff Slutsky—Bausch & Lomb
Mr. Steve Schaus—Sequa
Dr. Gerry Culkin and Mr. Les Birdwisa—Goodyear
Mr. Rick Moore—MSA
Dr. Bob Wolf—Andersen Windows
Mr. Rob Gier—Baxter
Mr. Terry Bunge and Dr. Ed Miller—ACIST Medical Systems,
  Bracco Diagnostics

And the world's newest professor of engineering to take all this crazy stuff to our young upstarts at Penn State University (my alma mater), Mr. James Hendrickson

You all have contributed to the ideals and discipline that are represented in this text. My thanks go out to each of you for standing up for doing the right things at the right time—for the right reasons....

Skip Creveling
August 2006

CONTENTS

Foreword by John Boselli   xiii

Preface   xv

About the Author   xxi

Chapter 1:   Introduction to Six Sigma for Technical
Processes   1

Setting the Stage for Growth   2
The Process Context   7
Phases and Gates   11
The Processes for Growth   15

Chapter 2:   Scorecards for Risk Management
in Technical Processes   21

Scorecards in Technical Processes   22
Checklists   23
Scorecards   23
Task Scorecards   26
Task Fulfillment vs. Gate Deliverable
Requirement   27
Gate Review Scorecards   29
Summary   32

Chapter 3:   Project Management in Technical
Processes   35

Six Sigma Contributions to Project
Management in Technical Processes   36
Designing Cycle-Time: Critical Paths of
Key Technical Tasks/Toolsets by Phase   38
Nine Key Steps   40
Modeling Technical Task Cycle-Time Using
Monte Carlo Simulations   42

Documenting Failure Modes in the Critical Paths of Technical Tasks by Phase   46

Integrated Project Plan   49

Chapter 4:   Strategic Product and Technology Portfolio Renewal Process   51

Six Sigma in the Strategic Product and Technology Portfolio Renewal Process   52

Process Discipline in Product and Technology Portfolio Renewal   55

The Phases of Product and Technology Portfolio Definition and Development   60

The IDENTIFY Phase of Product Portfolio Definition and Development   64

The DEFINE Phase Tools, Tasks, Deliverables, and Requirements   72

The EVALUATE Phase Tools, Tasks, Deliverables, and Requirements   81

The ACTIVATE Phase Tools, Tasks, Deliverables, and Requirements   86

Summary of the Major Steps for Product Portfolio Definition and Development Process   89

Chapter 5:   Strategic Research and Technology Development Process   95

Six Sigma–Enhanced Research and Technology Development   96

The I^2DOV Roadmap: Applying a Phase-Gate Approach to Research and Technology Development   100

I^2DOV Phase 1: Invent/Innovate Technology Concepts   103

I^2DOV Phase 1: Invention and Innovation Tasks   105

A General List of Phase 1 Tools, Methods, and Best Practices   116

I^2DOV Phase 2: DEVELOP Technology   120

A General List of Phase 2 Tools, Methods, and Best Practices   128

I²DOV Phase 3: Optimization of the Robustness
of the Baseline Technologies   132

A General List of Phase 3 Tools and
Best Practices   144

I²DOV Phase 4: Verification of the Platform or
Sublevel Technologies   147

A General List of Phase 4 Tools and
Best Practices   156

References   159

**Chapter 6:   Tactical Product Commercialization
Process   163**

*Six Sigma–Enhanced Product
Commercialization   164*

*Preparing for Product
Commercialization   165*

*Defining a Generic Product Commercialization
Process Using the CDOV Roadmap   168*

*The CDOV Process and Critical Parameter
Management during the Phases and Gates
of Product Commercialization   171*

Concept Phase: Develop a System Concept Based on
Market Segmentation, the Product Line, and
Technology Strategies   171

Design Phase: Design Subsystem-, Subassembly-,
and Part-Level Elements Based on System
Requirements   197

*The CDOV Process and Critical Parameter
Management during the Phases and Gates
of Product Commercialization   226*

Optimize Phase: Optimize Sublevel Designs
and the Integrated System   226

Verify Phase: Verification of Final Product Design,
Production Processes, and Service Capability   249

Chapter 7:   Fast Track Commercialization   275

*Six Sigma Applications for Fast Track
Commercialization Projects   276*

*DMAIC Six Sigma Project Capability to
Support a Fast Track Project   277*

Technology Development Capability to Support Fast Track Projects 279

Concept Phase Risk Profiles and Tool-Task Recommendations 280

Design Phase Risk Profiles and Tool-Task Recommendations 283

Optimize Phase Risk Profiles and Tool-Task Recommendations 285

Verify Phase Risk Profiles and Tool-Task Recommendations 287

Macrosummary of Fast Track Projects 289

Chapter 8: Operational Post-Launch Engineering Support Processes 293

Post-Launch Product, Production System, and Service Support Engineering Six Sigma in the Operational Technical Process 294

Hard vs. Easy Data Sets 301

The Tools, Methods, and Best Practices That Enable the LMAD Tasks 302

The LAUNCH Phase 305

The MANAGE Phase 307

The ADAPT Phase 308

The DISCONTIUE Phase 310

Chapter 9: Future Trends in Six Sigma and Technical Processes 317

Trends in Lean and Six Sigma for Technical Processes 318

Glossary 323

Index 351

FOREWORD

To survive in today's global market, companies must possess the capability to continually reinvent themselves in response to changing customer needs and aggressive worldwide competition. History has shown us that companies that were once icons in their respective fields can lose their competitive advantage.

To grow and prosper in this climate, companies must develop and sustain the capability to offer superior products and services that are aligned to existing and emerging customer needs. Companies must possess the capability to create value that customers are willing to pay for. Real leaders help to direct the evolution of their markets through insight, vision, and, most important, the ability to execute. The pressure is always on! Successful companies never sit back, for they recognize that there is always someone out there who wants their business.

In support of this mission, companies must possess robust, highly efficient new product development and commercialization processes. Obviously, this includes the capability to create robust and tunable technologies that can ultimately provide commercial value for the company. These aligned capabilities must work as an efficient, lean system to produce a steady, predictable stream of successful product launches. Many companies exhibit excellent subsystem performance but fail to achieve the ultimate goal due to a lack of coordinated planning and execution.

Effective business and technology leaders recognize the importance of properly managing this critical business process, for this is the lifeblood of the corporation. Leaders must see the whole picture, the end-to-end process, and understand the manner in which the pieces interact. Leaders are responsible for managing the company's scarce resources wisely to achieve business goals and objectives. Companies simply cannot afford to allocate resources to fixing newly released products or creating products that fail to garner market appeal.

This text provides an essential roadmap for business and technology leaders as they strive to achieve excellence in new product development. It discusses the role that DMAIC Six Sigma has played in process improvement, while clearly focusing on the application of lean Six Sigma thinking on the prevention of defects and the promotion of successful product launches. It provides a logical framework that leaders can adopt to promote excellence in all facets of new product development, including, but not limited to, market sensing, portfolio management, technology development, commercialization, and post-launch service and support.

In closing, this text is worth the read. It brings it all together for business and technology leaders. I want to thank C. M. Creveling and the PDSS team for creating this valuable tool. Well done!

John D. Boselli, PE
Vice President, Quality Management and Regulatory Compliance
BD Diagnostics, PreAnalytical Systems, Beckton Dickinson & Co.

PREFACE

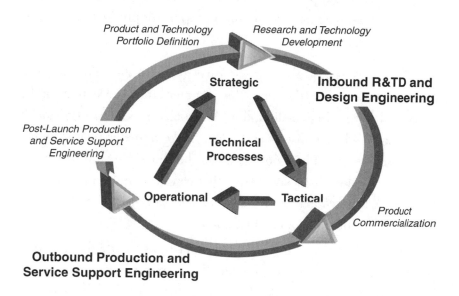

What Is in This Book?

This book is all about Six Sigma for technical leaders and management professionals and the processes they oversee. It is structured to be a guide for designing the flow of Six Sigma–enhanced work and measuring results within and across technical processes. The kind of Six Sigma we explore is relatively new; it is the form of Six Sigma that prevents problems within well-designed and structured technical processes. Its focus is four process arenas for enabling a business to attain a state of sustainable growth:

- **Strategic portfolio-renewal process:** product and technology portfolio definition and development

- **Strategic R&TD process:** basic research and technology development

- **Tactical design-engineering process:** product commercialization

- **Operational production and service support engineering process:** post-launch technical support for production and service engineering

This book is not a comprehensive guide to all technical tasks across an enterprise. It is about the portion of those tasks that can be enhanced by Six Sigma discipline. We focus on *what to do* (major tasks enhanced by toolsets) and *when to do it* (major phases within our processes) as leaders, not as doers. The "how" part, for "doers," is a very detailed body of knowledge that can be found, in part, in our text *DFSS in Technology and Product Development* (by Creveling, Slutsky, and Antis; Prentice Hall, 2003).

Anyone interested in how Six Sigma tools, methods, and best practices can enhance and enable these four process arenas will benefit from this new book. We believe this text will help guide the reader to structure a "lean" workflow for completing the right technical tasks using the right tools, methods, and best practices at the right time. So, yes, this book is all about "lean" Six Sigma for technical processes and the results this produces to enhance sustainable growth.

This book has almost nothing to do with the older form of Six Sigma known for its five-step DMAIC process for solving problems and cleaning up quality defects. DMAIC stands for five distinct problem solving steps: *Define* a costly problem, *measure* the process (take credible data) where the problem exists, *analyze* the data to define root causes, *improve* the process so that it meets requirements, and, finally, *control* the process to keep it "on target" using a control plan. Our focus migrates away from these simple five steps to the actual technical processes that are used to run a modern enterprise on a day-to-day basis.

Why did I write this book? To help expand the understanding of executives, leaders, and managers beyond the kinds of Six Sigma paradigms, workflows, measurement rigors, and "lean" process disciplines that exist in the world of Design for Six Sigma found in our first text, previously mentioned. My text on DFSS, which came out in 2003, has become a recognized resource for R&D and product commercialization engineers—*the doers*. Every time we teach and mentor engineering teams on DFSS and TDFSS (technology development for Six Sigma), they ask, "Do our leaders agree with this approach and are they going to support us? Are they aware that they are flooding us with too many projects? Do they realize that our ability to really do what we do best and do it right the first time is being compromised by giving us too many projects and too little time to complete our tasks? Shouldn't they be working in front of us to assure us we have what we need to do this right? Shouldn't they prevent problems on our behalf where they have the power and resources to do so?" The answer, of course, is always "Yes, yes they should."

This text will help you align with and proactively support your teams if you intend to integrate Six Sigma discipline into your enterprise workflow. This book is part of an exciting new set of books from Prentice Hall called the Six Sigma for Innovation & Growth Series. That means there are more books like it to help educate others in your leadership ranks as you all integrate Six Sigma where it adds value across the enterprise. If you want to see how inbound and outbound marketing relates to all of this in management and leadership roles, see our new text, *Six Sigma for Marketing Processes* (by Creveling, Hambleton, and McCarthy; Prentice Hall, 2006), written for marketing professionals.

DFSS and TDFSS (technology development for Six Sigma) are integrating to form a unified approach for those who are trying to improve product commercialization cycle time together. So this book is partly DFSS for R&TD professionals. We go well beyond just talking about product commercialization in this book, though. We set the stage for a comprehensive Six Sigma–enabled workflow for technical professionals that crosses the four process arenas we mentioned earlier: product and technology portfolio renewal, R&TD, commercialization (tactical inbound engineering), and product line management (operational outbound engineering). That is why the logo for this book looks the way it does. This text promotes two distinct themes:

1. Plan your product and technology portfolio well so you take the right number and order of projects into R&TD and commercialization.

2. Do the projects you have activated from your options very well within your R&TD and product commercialization processes.

Take a moment to reflect on the logo (it appears at the beginning of each chapter), and you will see our view of how engineering workflow is structured in the text.

The book consists of nine chapters. Chapter 1, "Introduction to Six Sigma for Technical Processes," lays out the whole integrated story of Six Sigma in technical processes. It covers the big picture of how all four technical process arenas work in harmony. One without the others is insufficient for actively sustaining growth in a business. This chapter also sets the stage for how phases and gates form a control plan for getting work done properly with minimal risk. Six Sigma is commonly associated with establishing a control plan; our control plan is the system of phases and gates used to structure the flow of work and assess data for risk management and decision making. We discuss how phase-gate systems are built and how they are "loaded" with tasks that can be enhanced and enabled with Six Sigma tools, methods, and best practices.

Chapters 2, "Scorecards for Risk Management in Technical Processes," and 3, "Project Management in Technical Processes," work closely together. Chapter 2 is about a system of integrated scorecards that measures risk accrual from tool use to task completion, to gate deliverables for each of the four technical processes. Chapter 3 gives a project management view of how technical teams can design and manage their work with a little help from some very useful Six Sigma tools (primarily Monte Carlo Simulations and Project Failure Modes and Effects Analysis [FMEA]). Chapter 3 will help you lean out your technical tasks and assess them for cycle-time risk. It can be used with traditional critical path methods or critical chain and buffering methods.

Chapters 4–8 contain more detailed views within each of the four technical processes. They lay out gate requirements and gate deliverables within phase tasks and detail the enabling tools, methods, and best practices that help technical teams complete their work on a timely basis. They offer a set of standard work (a lean term) that can be adaptively designed into your technical processes where you live on a daily basis. Chapter 7, "Fast Track Commercialization," is very unique; it discusses how to design a "Fast Track Project" for commercializing a high-risk, high-reward opportunity to help grow your business. A Fast Track Project is a faster-than-lean product-commercialization project that is rushed through your phase-gate process with some critical, value-adding tasks not fully completed. This rushed form of commercialization should be used sparingly and should not be the normal way you work. The use of Six Sigma in portfolio renewal, R&TD, and post-launch processes is discussed so that you can justify doing such a risky project during the phases of commercialization.

These chapters will help you design lean, Six Sigma–enabled work so you have efficient workflow and low variability in your summary results. The standard work in these chapters will help prevent problems and ultimately sustain growth. This is because what you do will add value and help ensure that your business cases reach their full entitlement. When business cases do what they promise, you will see sustained growth.

Chapter 8, "Operational Post-Launch Engineering Support Processes," wraps up everything quickly and succinctly. We know technical management professionals are very busy folks, so we try to get the right information to you in a few short chapters so you can help design work and lead your teams to new levels of performance as you seek to sustain growth within your business.

We would like to thank all the great people at Prentice Hall and those who helped with critical reviews prior to publishing for their support and hard work to make this part of the Six Sigma for Growth Series a success:

Prentice Hall staff:

Bernard Goodwin, Publisher

Michelle Housley, Editorial Assistant

Curt Johnson, Executive Marketing Manager

Heather Fox, Publicist

Christy Hackerd, Production Editor

Damon Jordan, Development Editor

Krista Hansing, Copy Editor

Sandra Schroeder, Cover Designer

Bill Camarda, Cover Writer

Nonie Ratcliff, Compositor

Angie Bess, Indexer

Critical review team:

Roger Forsgren

Paul Hellinga

David C. Trimble

William Tynes

ABOUT THE AUTHOR

Clyde "Skip" Creveling is president and founder of PDSS, Inc. Since its founding in 2002, Mr. Creveling has led the DFSS initiatives at Merck, Motorola, Carrier Corporation, StorageTek, Cummins Engine, Beckton, Dickinson & Co., Mine Safety Appliances, and Callaway Golf. Before founding PDSS, Mr. Creveling was an independent consultant, DFSS Product Manager, and DFSS Project Manager with Sigma Breakthrough Technologies, Inc.(SBTI). During his tenure at SBTI, he served as the DFSS Project Manager for 3M, Samsung SDI, Sequa Corp., and Universal Instruments.

Mr. Creveling was employed by Eastman Kodak for 17 years as a product-development engineer within the Office Imaging Division. He also spent 1 1/2 years as a systems engineer for Heidelberg Digital as a member of the System Engineering Group. During this period, he worked in R&D, product development/design, system engineering, and manufacturing. Mr. Creveling has five U.S. patents.

He was an assistant professor at Rochester Institute of Technology for 4 years, where he developed and taught undergraduate and graduate courses in mechanical engineering design, product and production system development, concept design, robust design, and tolerance design. Mr. Creveling is also a certified expert in Taguchi methods.

He has lectured, conducted training for, and consulted on Product Development Process improvement, design for Six Sigma methods, technology development for Six Sigma, critical parameter management, robust design and tolerance design theory, and applications in numerous U.S., European, and Asian locations. He has been a guest lecturer at MIT, where he assisted in the startup of a graduate course in robust design within the Master of Science in System Development and Management program.

Mr. Creveling coauthored *Engineering Methods for Robust Product Design* (Addison-Wesley, 1995, ISBN 0-201-63367-1) and the world's first comprehensive text on developing tolerances, *Tolerance Design* (Addison-Wesley, 1997, ISBN 0-201-63473-2). This book focuses on analytical and experimental methods for developing tolerances for products and manufacturing processes. He also is the coauthor of two new texts for Prentice Hall's Six Sigma for Innovation and Growth Series, *Design for Six Sigma in Technology and Product Development* (2003, ISBN 0-13-009223-1) and *Six Sigma for Marketing Processes* (2006, ISBN 0-13-199008-X), and a Digital Shortcut titled *What Is Six Sigma for Technical Processes?* (2006, ISBN 0-13-157422-1).

Mr. Creveling is the editorial advisor for Prentice Hall's Six Sigma for Innovation and Growth Series.

He holds a Bachelor of Science degree in mechanical engineering technology and a Master of Science degree from Rochester Institute of Technology.

1

INTRODUCTION TO
SIX SIGMA FOR
TECHNICAL PROCESSES

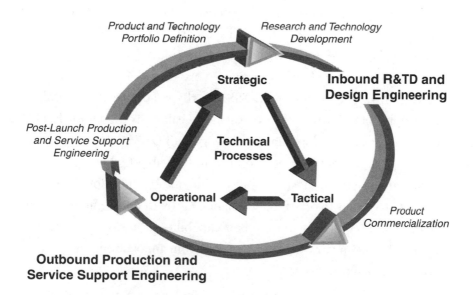

Setting the Stage for Growth

Product and technology portfolio planning, research and technology development (R&TD), product commercialization, post-launch production and service support engineering, and Six Sigma ... is there a link? Can technical professionals successfully integrate the tools, methods, and best practices of Six Sigma into their day-to-day work? The answer is, yes—but you have to understand how Six Sigma is evolving in technical processes to make the connection. The original 13 or so tools of Six Sigma grow considerably when your work is focused on preventing problems instead of reacting to them. Certainly, design and manufacturing engineering professionals have proven that Six Sigma works well in their processes. DFSS has become an accepted approach to enhancing the capability to commercialize new products. The DMAIC process has been used to improve existing designs and the processes that make them in many companies across the world.

A fairly large number of companies use DFSS in their day-to-day commercialization activities, called standard work, to enhance design team performance and provide optimal production and service support engineering. A few companies—very few—have a thriving culture of using a form of DFSS in their advanced technology development and research processes. We call it Six Sigma for R&TD or Technology Development for Six Sigma (TDFSS). The work done in this environment produces robust and tunable technologies: platforms and modular designs that enable multigenerational product families to reach their strategic financial growth entitlement (potential). This discipline enhances their capability to maximize growth in terms of ongoing revenue generation from innovation within their core technical processes.

In this book, technical processes include technology portfolio architecting, R&TD, product commercialization, and post-launch engineering work. Two distinct categories now exist in which you can classify Six Sigma applications for technology portfolio architecting, R&TD, product commercialization, and post-launch engineering. The older category involves using DMAIC Six Sigma and Lean methods to correct problems and increase flow in existing technical processes—quick, "emergency" action. The new category involves enabling and enhancing technical processes to prevent problems before they become an issue. Our goal is to use Six Sigma on a sustained basis to become consistent and predictable at conducting value-adding tasks during technology portfolio architecting, R&TD, commercialization, and post-launch engineering support.

The shift in the Six Sigma paradigm that this book addresses comes from reacting to problems created by rushing technologies into design and products into the market, to prevent problems, by design, through proactively planning work. Paraphrasing what Albert Einstein has stated, "We cannot solve our problems by approaching them with the same state of mind we had when we first created them." That is why many see "design" for Six Sigma as just the beginning of the shift from a react-strategy to a prevent-strategy. Sometimes it is worth the risk of rushing a commercialization project, but these are the exceptions, not the rule. This book shows, as Paul Harvey says, "the rest of the story."

Problem prevention in technical processes can be done during "inbound" R&TD and product design engineering, as well as "outbound" post-launch engineering processes.

Inbound R&TD is focused on strategic technology portfolio definition, development, optimization, and transfer. Inbound product design engineering is focused on tactical product commercialization to rapidly prepare a new design, which often possesses transferred, new technology to fulfill launch requirements.

Outbound post-launch engineering is focused on operations in post-launch production and service engineering support. The service engineering professionals often exist as a "reactionary force" to fix things when they go awry (like Mr. Wolf in the movie *Pulp Fiction*).

Our focus is on planning engineering changes and upgrades in this environment when we want to increase margins. We use DMAIC Six Sigma when things blow up and a crisis ensues. We plan to help the company grow using the post-launch engineering resources when they are not putting out fires.

The vocabulary of this book breaks technical functions and processes into inbound growth enablement and outbound growth enablement arenas. Everything we discuss is primarily about enabling growth at the top line instead of using emergency defect reduction and bottom-line financial-control issues. We can use DFSS reactively to cut costs and improve product designs and production processes— but that is not how you should structure technical processes for sustainable growth. We want to see very few cost-cutting and problem-solving projects in the future world of Six Sigma–enhanced work. We want to structure technical work processes to prevent problems so that we mainly work on creating new products, not fixing old ones.

Some of this "fixing" will always need to be done—but not a lot. We saw one company devote more than 75 percent of its engineering workforce to working on corrective actions for designs that were launched in the marketplace. That is simply unacceptable use of advanced technology and design-engineering resources. That kind of behavior can damage your brand. Shipping products before they are properly optimized is done all the time, but that is no way to approach sustainable growth. Companies that launch products prematurely and then fix them do get bursts of market share, but they almost never show sustainable growth. Just look at the performance of the U.S. vs. Japanese auto industry, and you will see a sobering example of this. Why does everybody want to be "like" Toyota? Being like Toyota is almost impossible for a Western company functioning in a Western culture. The closest we are likely to get to being like Toyota is to selectively integrate some of the things found in this book that can add value to the way we develop and consume data. We can't give you your specific recipe; you must custom-develop that with the help of this text. Toyota is unique, and that company's recipe won't work for you. Many things in this book Toyota does very well—by habit. Other things in here would totally screw up the company because of its unique culture. Businesses are a little like our human bodies: Some drugs cure many of us, yet the same drug kills a small percentage of us. We have to look hard at exceptions.

The focus of this book is on the new frontier of applying Six Sigma and Lean discipline to an integrated, enterprise Six Sigma strategy that creates measurable capability in sustaining top-line growth. "Old" Six Sigma is all about fixing problems and implementing bottom-line cost control; we are on a different financial journey, seeking top-line growth. Controlling cost is important, but creating sustainable growth is even more important. Growth is based upon investment in strategic projects that generate new revenues and increase margins.

The use of Six Sigma to control bottom-line costs has usually been associated with manufacturing and supply-chain processes, environments in which very large costs and inefficiencies have traditionally built up in most companies. To be sure, R&TD and product design engineering organizations can create and incur significant costs and waste in their own right. Excessive costs arise while R&TD and product design engineering organizations are attempting to develop, transfer, and commercialize a technology portfolio for top-line growth. Applying Lean and DMAIC Six Sigma to their process design (phase-gate structure) can fix these inefficiencies.

Six Sigma in technical processes, as presented in this book, enables the structuring and application of technical tasks and tools needed to develop growth—after the waste and process inefficiencies have been defined, measured, analyzed, improved, and controlled. So the value-add from this book is to define how Six Sigma applies *after* a DMAIC project has readied a technology-development or product-commercialization process to actually be used on a day-to-day basis.

This book will be quite helpful in the *Improve* phase of a DMAIC phase-gate redesign project. This book has suggestions for standard work that you can do every day as you go through your phases and gates processes. When you have strong, stable, well-designed R&TD and commercialization processes, you can apply designed Six Sigma toolsets to enhance the predictability and quality of day-to-day engineering tasks. You can then run numerous Six Sigma–enhanced growth projects through your processes with an increased likelihood of sustaining growth.

Inbound R&TD work can cause problems and waste by under-developing the right platforms and modular designs needed to enable the proper commercialization of new families of products and their supporting production systems. We can develop and transfer the wrong technologies or immature technologies within the mix of products and, hence, miss the growth numbers promised in the business cases that were supposed to support the long-term financial targets of the company.

Outbound post-launch engineering creates problems and waste by failing to develop the right data to make key decisions about managing, adapting, and discontinuing the various technical elements of the existing product or processing lines (manufacturing, assembly, packaging, and servicing). It also fails to get the right information back upstream to the product portfolio and technology-renewal teams so they can renew the portfolios based upon real, up-to-date data and lessons learned from customer feedback and the technical support experts in the field.

Before we look to the future of growth-oriented Six Sigma for technical processes, let's build a little history of how Six Sigma corrects problems to then reduce costs. If a technical process is broken and inefficient, we need to correct that condition. Then we can move on to enable that process to focus on business growth by the tasks we choose to do (and not do) on a day-to-day basis.

The Process Context

Six Sigma started out as a problem-solving process. It was built primarily on the foundations of Shewhart's statistical process control steps and Total Quality Management tools and methodologies. The steps Six Sigma practitioners use to solve problems are abbreviated as DMAIC:

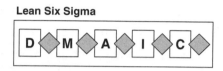

1. **Define** the problem.

2. **Measure** the process and gather the data that is associated with the problem.

3. **Analyze** the data to identify a cause-and-effect relationship between key variables.

4. **Improve** the process so the problem is eliminated; measured results meet requirements.

5. **Control** the process so the problem does not return—or, if it does, it is controllable.

The DMAIC process is easy to learn and apply. It provides strong benefits to those who follow its simple steps using a small, focused set of tools, methods, and best practices. Companies are successful in using this approach because they train small teams to stick to this approach without wavering in their completion of specific tasks within projects that typically last 6 to 9 months. They learn the DMAIC process and its vocabulary, and apply the tools much like a well-trained surgical team conducting an operation. They are focused, enabled by their project sponsors, and absolutely deliver on their documented requirements and specific goals. The key elements in a DMAIC project are team discipline, structured metrics, toolsets, and execution of a well-designed project plan that has clear goals and objectives. When large numbers of people across a multinational company use the simple steps of DMAIC Six Sigma to solve problems the same way everywhere using a common vocabulary, objectives— result targets are much harder to miss. If everyone solves problems differently, nonsystematic, global process-improvement initiatives break down. Without process discipline, cost and waste reduction are often haphazard and impossible to integrate and report at the corporate level. In an undisciplined environment, cost reduction and control are unsustainable and unpredictable. Sometimes it works, but usually the projects fizzle.

If a process cannot be improved as it is currently designed, another well-known, reactive Six Sigma approach can be applied. The DMADV process is used to fundamentally redesign a process when the existing process cannot be improved using the standard DMAIC process and tools.

DMADV stands for this approach:

- **Define** the problem and the opportunity a new process represents.

- **Measure** the process and gather the data that is associated with the problem, as well as VOC data associated with the opportunity to design a new process.

- **Analyze** the data to identify a cause-and-effect relationship between key variables, generate new process concepts, and select a new process architecture from numerous alternatives.

- **Design** new, detailed process elements and integrate them to eliminate the problem and meet the new requirements.

- **Validate** the new process so you can meet the new process requirements.

Some have defined another set of steps for this reactive, process redesign approach, called the DMEDI approach. This is essentially the same kind of approach as DMADV, but a different vocabulary and added tools from Lean methodology. DMEDI stands for this process:

- **Define** the problem or opportunity.

- **Measure** the process and gather the data that is associated with the problem or opportunity (VOC and requirements).

- **Explore** the data to identify a cause-and-effect relationship among key variables, requirements, and conceptual alternatives, arriving at one concept to take into detailed design.

- **Design** a new process so the problem is eliminated and measured results meet requirements.

- **Implement** the new process under a control plan.

Whether you use DMADV or DMEDI, the goal is to design a new process to replace the incapable, existing process. This is still classic, "old-style" Six Sigma for problem solving. We mean no disrespect when using the term "old-style;" we are merely trying to define the future of Six Sigma and, by necessity, have to draw a distinction between what is old and what is new.

Six Sigma discipline, tool-task linkage, project structure, and, most important, result metrics are migrating away from problem solving. The new frontier for Six Sigma is in problem prevention. As they say, "an ounce of prevention is worth a pound of cure." Problem prevention has to occur in our daily workflow to be effective. DMAIC and DMADV/DMEDI are needed to get processes improved and under control. When this is finished, another form of Six Sigma–enabled process execution is required. Within technical processes, Six Sigma is a set of structured tools/tasks/deliverables for problem prevention during the phases and gates of R&TD, product commercialization, and post-launch engineering processes. So we don't use DMAIC or any of its derivatives as names for our technical phases in our standard business process culture. Bob Cooper, PRTM, and their peers have dominated the conventional structuring and naming of standard phase-gate processes. This book seeks to build on their work.

To summarize, the process environments that technical professionals work in on a day-to-day basis is not a DMAIC-, DMADV-, or DMEDI-based workflow structure. The work of technical professionals breaks down into three process arenas:

- Strategic technology portfolio planning, development, and transfer (renewing, developing, and transferring the technology portfolio)

- Tactical product commercialization (commercializing a specific product)

- Operational production and service engineering (supporting the launched product portfolio)

Each of these arenas has a flow of repeatable work, a process context that is quite different than DMAIC or DMADVI/DMEDI. The work is made up of specific tasks that are enabled by flexible, designable sets of tools, methods, and best practices. These sets of tasks and enabling tools are not the steps of DMAIC or DMADV/DMEDI. They align with phases that can be designed to prevent problems, to limit the accrual of risk and enable the right kind and amount of data to help make key decisions at your gate reviews. So DMAIC and DMADV/DMEDI approaches help get your processes improved, redesigned, and under control when you have problems—but they are not the steps used to develop product portfolios, develop new technologies, commercialize products, or support product lines. Those processes follow a different set of "steps" that we call phases. We know you use your own company's vocabulary for these phases, so for the purposes of communication, we have given generic names to phases in this text; we suggest that you use your own vocabulary and just use ours as "word bridges" to help understand what we are discussing in this book.

Phases and Gates

A phase is a period of time during which value-adding tasks are

- Enhanced by tools, methods, and best practices

- Applied in a planned and disciplined manner

- Used to deliver very specific deliverables that fulfill a predetermined set of gate requirements

A gate is a stopping point within the flow of the phases within a given project to

- Assess risk accrued from past results

- Make decisions about future work relative to clear requirements

Gate reviews are greatly enhanced by the predevelopment of clear requirements and the deliverables that fulfill them. This is the responsibility of a balanced team of gatekeepers. Making up requirements as you go is a sure way to corrupt cycle-time and inhibit the potential to grow on a sustainable basis. The most common malady we see today in modern management style is to use unstructured gate reviews, poorly attended by a partial team of managers, to assess incomplete deliverables and then use spur-of-the-moment judgment to make key decisions. The deliverables are incomplete because the management team has chosen to rush the teams through a flow of work that they have not taken the time to preassess for risk. The risk builds from tasks that are poorly designed and partially completed, along with a lack of data integrity stemming from improper use of tools, methods, and best practices that are known to enhance the effectiveness of the tasks.

Our graduate schools of business management are failing to prepare our leaders to be effective "gatekeepers." We know this to be true because we have asked hundreds of gate-keeping managers how they were prepared to do this key management task. Not one replied that they had adequate training and preparation to be a competent gatekeeper by any measure or standard of management excellence.

Two specific things occur at this stopping point:

- A thorough assessment of deliverables is conducted, relative to their predetermined requirements, as a result of the tasks and their enabling tools, methods, and best practices that were conducted during the previous phase (specifically, the high-risk areas that did not fully fulfill the gate requirements).

- A thorough review of the project management plans is completed (typically in the form of a set of PERT charts, Monte Carlo simulations, and critical path FMEAs) for the *next* phase of work. Gate reviews mostly focus on preventive measures to keep flaws and mistakes from occurring in the next phase. The integrity of data is the basis on which risk is assessed and decisions are made. Tool-task-deliverable design and linkage is the foundation of fulfilling requirements. A good gate-keeping team is always one phase ahead of the teams doing the work, from a problem-prevention perspective. Gates are also used to manage work-around plans, to keep things moving forward after difficulties and problems have been identified by using data or the lack thereof. A well-designed gate review takes into account that things will go wrong and enables the team to deal with these inevitabilities. Even a complete, properly trained gate-keeping team can't think of everything that can go wrong, but that is no excuse for not trying.

It is to be expected that some elements of a project will go through a gate on time, while other parts could be experiencing unexpected difficulties in meeting the gate requirements. It is common for project teams to use a three-tiered system of colors to state the readiness of their data and summary deliverables to pass through a gate. We cover this in detail in Chapter 2, "Scorecards for Risk Management in Technical Processes."

- Green means the design has passed the gate criteria with no major problems.

- Yellow means the design has passed some of the gate criteria but has numerous (two or more) moderate to minor problems that must be corrected for complete passage through the gate. A clear resolution path is documented.

- Red means the design has passed some of the gate criteria but has one or more major problems that preclude the design from passing through the gate. No known resolution path is documented.

A resolution path is a set of tasks and the enabling toolset that will be used to correct problems or shortfalls in deliverables relative to their requirements. This corrective path has a predicted timeline and a promised date of completion that takes into account that bad things can happen to good people while they work. Most gate problems occur when the people who do the work are rushed. The problems also stem from incomplete or ambiguous requirements. If you make a "best practice" out of rushing your people and giving them unclear requirements, they will not complete their critical tasks completely and will not use a balanced set of tools to ensure the quality of their work. You can expect a lot of "design and marketing scrap and rework." This is a major root cause of poor cycle-time.

Gate passage is granted to projects whose deliverables are Green. Gate passage is provisionally granted to designs that are Yellow. Gate passage is denied to designs that are Red. Time and resources are planned to resolve the problems (it is not unusual to see the time ranging between 2 and 6 weeks and, in some cases, longer). Workaround strategies are put in place to keep the whole project moving forward while the design elements that have failed gate passage are corrected. This situation requires engineering managers to develop phase-transition strategies that allow for *early and late* gate passage. In the ideal state, all design elements pass all the gate criteria at the same time for every gate. Of course, this almost never happens. For complex projects, it is unrealistic to expect anything other than a series of "unexpected" problems. For the engineering manger thinking preventively, the second order of business in developing a phase/gate commercialization process is to include a phase-transition strategy and policy that effectively deals with managing the unexpected. The *first* order of business is structuring each phase with a workflow of tasks enabled by a balanced portfolio of tools and best practices that prevents unexpected design problems. This is done by focusing on the tools and best practices that use *forward-looking* metrics that measure functions (scalar and vector-engineering variables) instead of counting quality attributes (defects). The key is to prevent the need to react to quality and cycle-time problems.

It is nearly universally accepted that using an overarching product-development process to guide technology development and product commercialization is the right thing to do. Often overlooked is the value of a broad, balanced portfolio of tools and best practices that is linked to key tasks that are scheduled and conducted within each phase using the methods of project management.

When portfolio renewal, R&TD, commercialization, and post-launch teams structure "what to do and when to do it" using PERT charts under the guiding principles of project management, they have their best chance of meeting cycle-time, quality, and cost goals. If they leave the process open to personal whims and preferences, chaos will soon evolve and the only predictable things will become schedule slips, the need for design scrap and rework, and perhaps "career realignment" for the culpable managers (if they are actually held accountable for their mismanagement of the company's resources!).

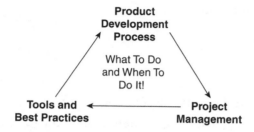

FIGURE 1.1   Triangle of process excellence.

The Processes for Growth

In the strategic process environment, we have identified two generic processes segmented into four distinct phases. The first is the IDEA process for product and technology portfolio definition and renewal:

Product Portfolio Renewal

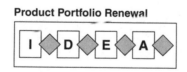

- **Identify** phase
- **Define** phase
- **Evaluate** phase
- **Activate** phase

The I^{2}DOV process for technology development and transfer to commercialization involves these phases:

Technology Development

- **Invent/Innovate** phase

- **Develop** phase

- **Optimize** phase

- **Verify** phase

In the Tactical product design-engineering process environment, we have identified a generic product-commercialization process segmented into four distinct phases:

The CDOV process for specific product-commercialization projects involves the following:

Product Commercialization

- **Concept** phase

- **Design** phase

- **Optimize** phase

- **Verify** phase

In the operational product and production support engineering process environment, we have identified a generic process segmented into four distinct phases:

The LMAD process for managing the portfolio of launched products involves these phases:

Production and Service Support

- **Launch** phase

- **Manage** phase

- **Adapt** phase

- **Discontinue** phase

Each of these processes possesses distinct phases in which designed sets of tasks are completed. Each major task can be enabled by one or more tools, methods, or best practices that give high confidence that the technical team is developing the right data to meet the task requirements for each phase of standard work. It has become common to assess results and define risk at the end of a phase by holding a gate review. Gate reviews are an important part of risk management and decision making for technical executives and engineering management professionals.

This book focuses on the integration of technical process structure, requirements and deliverables (phases and gates for risk management), project management (for design and control of technical task cycle-time), and balanced sets of technical tools, methods, and best practices. The new model for process integration of Six Sigma looks like Figure 1.2.

It is all about integrating Six Sigma tool-task-deliverable-requirement flows within and across your business processes (see Figure 1.3).

This is the context we need to prevent problems within and across the strategic, tactical, and operational technical processes after they have been designed properly.

Again, if your technical processes are broken, incapable, or out of control, use DMAIC or DMADV/DMEDI approaches to improve or redesign them. This book is useful when such projects get to the improve or redesign steps. This book answers the question of what to do and when to do it within well-structured technical processes. See *Winning at New Products*, by Robert Cooper (ISBN 0-7382-0463-3), or *Setting the PACE in Product Development*, by Michael McGrath

(ISBN 0-7506-9789-X), to help in process-architecting a technical phase-gate process. Six Sigma in technical processes is really focused on functional excellence within a structured phase-gate context. This approach is an enhancement to your phase-gate processes and the detailed flow of work inside them.

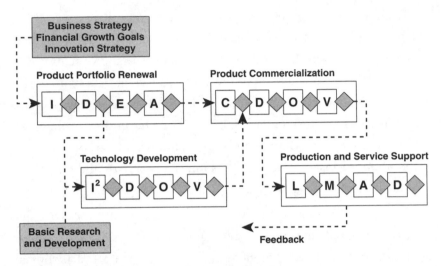

FIGURE 1.2   **Integrated process flow diagram.**

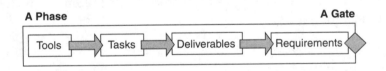

FIGURE 1.3   **Phase-gate linkage diagram for tools, tasks, deliverables, and requirements.**

Technical processes and their deliverables must be designed for efficiency, stability, and, most important, measurable results. We work within the IDEA, I²DOV, CDOV, and LMAD processes, applying their accompanying tool-task sets to create measurable deliverables that fulfill the gate requirements. You can choose to call your process phases by different names—that's fine. It's what you do and measure that really matters.

A common malady we find in most companies is that they are "deliverable" focused, not requirement focused. When deliverables drive your processes, you tend to get mired in build-test-fix cycles as management compromises its way to consensus while pondering the risk associated with the current state of deliverables. With a requirement-based process, teams have boundaries and targets to hit, and deliverables are constrained. You know precisely when to stop doing certain flows of work. Deliverable-based processes tend to have ambiguous requirements. Projects and their deliverables can drift through endless committee reviews that really slow things down. Focus on developing clear, specific requirements, and link your deliverables to them; then it is easy to define tasks and tools to get things done efficiently.

What about using Six Sigma and lean methods on a project that is so "hot" that if we don't put it through our phases and gates at an extremely accelerated rate, we will miss a major opportunity for growth? We call this type of high-risk, high-reward commercialization project a Fast Track Project. These projects should be few in number and should be justified and underwritten by solid, well-designed, unrushed work in the portfolio renewal (IDEA) and R&TD (I^2DOV) processes. Yes, you can design a truncated, risk-assessed, and balanced flow of tasks and tools to rush a project. Chapter 7, "Fast Track Commercialization," describes, in general, how to structure such a project in the context of Six Sigma and lean methodologies.

To summarize this introduction, Six Sigma for technical processes is a relatively new approach to enable and help sustain growth. It is part of the solution. It contributes to the likelihood of a bright future offered by adapting Six Sigma to the growth arena. The linkage of Six Sigma tool/task clusters to R&TD and design tasks in strategic, tactical, and operational processes is where Six Sigma discipline adds value to engineering team performance. R&TD, design, and manufacturing engineering professionals can custom-design what to do and when to do it to fit these critical technical process arenas. Why? To manage risk and make sound, data-driven decisions based upon clear requirements as they seek to grow the company on a day-to-day basis.

2

SCORECARDS FOR RISK MANAGEMENT IN TECHNICAL PROCESSES

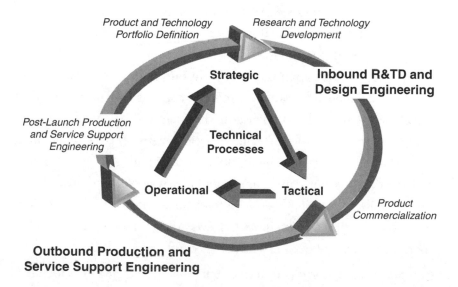

A system of scorecards to measure the use of tools, completion of tasks, and the resulting deliverables across strategic, tactical, and operational technical processes.

Scorecards in Technical Processes

Whether we are working in the strategic, tactical, or operational technical environment, we need to take into account how we will measure a team or individual's progress against goals. Accountability within and across technical teams is essential for proper risk management within a system of phases and gates. As technical work is designed and performed, the issue of pay for performance must also be addressed. Thus, we have two distinct reasons to establish formal methods for measuring technical task performance:

1. Manage risk and make key decisions at gate reviews and key project milestones

2. Pay for performance in light of specific requirements and deliverables

We all expect to be paid fairly for the value we add to the business from our work. We all want the technologies and products we are working on to be successful, and we want our compensation to be aligned with that success. But how do we pay fairly for projects that are unsuccessful for good reasons? The reasons become "good" and are justified based upon the data from our use of tool/task clusters. Without a balanced system of scorecards, it is not possible to tell whether we are truly compensating in a fair and balanced way, even when we cancel projects. The questions "Am I personally doing okay?" and "Are we, as a team, doing okay?" and "Are we all on target for meeting our gate requirements by way of our deliverables?" must be asked frequently enough to ensure that we can make adjustments when the data suggests that we are off-track relative to clear requirements. A system of scorecards can help.

Scorecards exist all around us, but what do they mean to technical professionals?

Checklists

One traditional form of accountability is the checklist. A checklist describes items that are assessed for two states of completion: done or not. Did you do a certain task? Did you use this tool or that tool to enable the task? Yes or no? Checklists suffer from a lack of discrimination; they reveal little regarding the quality or quantity of a tool used or how a task was done in terms of percent completion against its original requirements.

Scorecards

We use this hierarchy in this text to discuss accountability for getting the right things done at the right time:

1. Gate or milestone requirements

2. Gate or milestone deliverables

3. Tasks

4. Tools, methods, and best practices

This four-level flow is well-suited to measurement by using an integrated system of scorecards. A system of scorecards can be designed and linked so that each subordinate level adds summary content up to the next level.

The most basic level of scorecard is filled out by technical team members, who actually use specific sets of tools, methods, and best practices to help complete their within-phase tasks. This is called a tool scorecard. Tool scorecards are easy to fill out and should not take more that 15 minutes or so to complete at the end of a tool application. Typically, they are filled out in collaboration with the team leader for supervisory buy-in. When finished using a certain tool, the

person or set of team members who apply that tool should account
for the measurable items in Table 2.1.

TABLE 2.1   Tool Scorecard

SS-R&TD Tool	Quality of Tool Use	Data Integrity	Results vs. Require-ments	Average Score	Data Summary, including Type & Units	Task Require-ment

Quality of tool use can be scored on a scale of 1–10 based upon
the suggested criteria and levels in Table 2.2. You can adjust these
ranks as you see fit for your applications.

TABLE 2.2   Quality of Tools

Rank	Right Tool	Fullness of Use	Correct Use
10	x	High	High
9	x	Medium	High
8	x	High	Medium
7	x	Low	High
6	x	Medium	Medium
5	x	Low	Medium
4	x	High	Low
3	x	Medium	Low
2	x	Low	Low
1	Wrong tool		

The integrity of the data produced by the tool's use can be scored
using the suggested but modifiable ranking structure in Table 2.3.

TABLE 2.3   Integrity of Data

Rank	Right Type of Data	Proper Units	Measurement System Capability	Percentage of Data Gathered
10	Excellent	Direct	High	High %
9	Excellent	Direct	High	Medium %
8	Excellent	Direct	Medium	High %
7	Good	Close	High	High %
6	Good	Close	Medium	Medium %
5	Good	Close	Medium	Low %
4	Weak	Indirect	Medium	High %
3	Weak	Indirect	Low	Medium %
2	Weak	Indirect	Low	Low %
1	Wrong	Wrong	None	—

You can adjust the nature of the scoring criteria as you see fit for your applications. The key is to make very clear delineation among various levels of measurable fulfillment of the criteria.

The capability of the tool results to fulfill the task requirements is scored with the help of the following criteria:

10 = Results deliver all data necessary to completely support the fulfillment or lack of fulfillment of the task requirements.

9–8 = Results deliver a major portion of the data necessary to support the fulfillment or lack of fulfillment of the task requirements.

7–4 = Results deliver a moderate amount of the data necessary to support the fulfillment or lack of fulfillment of the task requirements.

3–1 = Results deliver a very limited amount of the data necessary to support the fulfillment or lack of fulfillment of the task requirements.

Here we want to account for how well our data fulfills original requirements. It is acceptable to find that, through a full set of data, we cannot meet the requirement a task was designed to fulfill. We have to reward technical professionals for doing good work that, unfortunately, tells us the truth about bad results. The intent is to avoid false positives and false negatives when making decisions about a project's viability. This metric helps quantify the underdevelopment of data and facts that can lead to poor decisions.

Task Scorecards

Task scorecards (see Table 2.4) have the capability to discriminate at the aggregate level of tool completion and summary performance relative to each major task and its requirements.

TABLE 2.4   Task Scorecard

Phase Task	Average Tool Score	Percentage Task Fulfillment	Task Result vs. Delinquent Requests	Red	Yellow	Green	Deliverable Requirements

A very insightful metric is the percent of completion that has been attained for each major task within a phase. We believe this is where the real mechanics of cycle-time are governed. If a technical team is overloaded with projects or is not given enough time to use tools, it is almost certain that it will not be able to fully complete its critical tasks. Undercompleted tasks are usually a leading indicator that a schedule likely will slip. Too few tools are being used, and those that are being used are not being fully applied. So a double effect results: poor tool use leading to incomplete data sets and tasks that simply are not finished. This means we make risky decisions on the wrong basis. The data is underdeveloped. High-risk situations are acceptable in our projects, but not because we are too busy to do things right. Task incompletion is a major contributor to why we make mistakes and fail to grow on a sustainable basis. This is a suggested ranking scale to illustrate how you can assign a value from 1 to 10 to quantify the level of risk inherent in the percent of uncompleted tasks:

10 = The task is complete in all required dimensions. A well-balanced set of tools has been fully applied to 100% completion.

9–8 = The task is approximately 80% to 90% complete. Some minor elements of the task are not fully done. A well-balanced set of tools has been used, but some minor steps have been omitted.

7–4 = The task is not complete somewhere across the range of 70% to 40%. Moderate to major elements of the task are not done, and tool selection and use has been moderate to minimal. Selected tools are not being fully used, and significant steps are being skipped.

3–1 = The task is not complete across a range of 30% to 10%. Very few tools have been selected and used. Steps have been heavily truncated, and major steps are missing altogether.

Task Fulfillment vs. Gate Deliverable Requirement

If we do complete all the critical tasks and use a balanced set of enabling tools to underwrite the integrity of our deliverables, we are doing our best to control risk. We can produce outstanding deliverables, full of integrity and clarity, that simply do not meet the requirements for the project and its business case. We then have great data that tells us we can't meet our goals. This is how a gate-keeping team can kill a project with confidence. Not many companies kill projects very often, and even fewer do it with tool/task/deliverable confidence.

We have two views to consider when managing risk and making gate decisions:

1. Fulfillment of the requirements so that we have a positive "green light" to continue investing in the project

2. Lack of fulfillment of the gate requirements so that we have a negative "yellow or red light" that signals a redirection of the project or an outright discontinuance of the project

Gatekeepers concern themselves with only yellow and red deliverables that they can do something about in terms of their influence, budgets, and control of resources. Gatekeepers should see only items of risk that they truly can affect. All subordinate areas of risk should be dealt with at the appropriate level. Bringing risk items to executives at the tool and task levels is a waste of their time; those items belong down in the functional management area, with those who have the ability to directly deal with the issue.

A color-coded scheme of classifying risk can be defined as follows:

- **Green**—All major deliverables are properly documented and fulfill the gate requirements. A few minor deliverables might be lagging, but present no substantive risk to the success of the project in time, cost, or quality.

- **Yellow**—A very few major deliverables are not complete or fall short of fulfilling their requirements. A corrective action plan is documented, and a very high probability exists that the problems can be overcome in a reasonable and predictable amount of time.

- **Red**—One or more major deliverables are not done or do not meet requirements, and no corrective action plan exists to close this gap. The project will be either killed or redirected or postponed until a corrective set of specific actions is defined and a predictable path to project timing is in hand.

This is a suggested set of ranking values to quantify the risk associated with varying levels of mismatch between what is delivered from a major task against the gate requirement:

10 = Results deliver all data necessary to completely support the fulfillment or lack of fulfillment of the gate requirements.

9–8 = Results deliver a major portion of the data necessary to support the fulfillment or lack of fulfillment of the gate requirements.

7–4 = Results deliver a moderate amount of the data necessary to support the fulfillment or lack of fulfillment of the gate requirements.

3–1 = Results deliver a very limited amount of the data necessary to support the fulfillment or lack of fulfillment of the gate requirements.

We have demonstrated how a tool scorecard documents the quality of tool use, integrity of the data, and fulfillment of a task. We have gone on to the next level of scoring risk by defining a task scorecard. Here we can quantify how well one or more enabling tools have contributed to completing a major task, what percentage of the task has been completed, and how well the deliverables from the task fulfill the gate requirements. We are now ready to look at the final summary scorecards that a gate-keeping team can use to quantify risk accrual at the end of a phase of work.

Gate Review Scorecards

Gate review scorecards help assess accrued risk and make decisions at each gate review or major milestone. Two kinds of gate reviews exist: functional-level gate reviews and executive-level gate reviews.

Functional reviews are detailed and tactical in nature and help prepare for executive review. Executive reviews are strategic in nature and look at macro-level risk in terms of how this particular project contributes to the overall portfolio of projects being commercialized or managed in the post-launch environment. Functional gatekeepers worry about microdetails within their particular project; executive gatekeepers worry about accrued risk across all projects that represent the future growth potential for the business as a portfolio. One looks at alignment of project risk across the business strategy, whereas the other looks at alignment of risk within the specific project's tactics and requirements. Functional reviews can be done for technical teams as independent events from other teams on the project. Executive reviews are summary presentations that include both technical and marketing, as well as all other forms of macro-gate deliverables.

Table 2.5 is an example of a generic template for a functional gate review.

Table 2.5   Functional Gate Review Scorecard

Gate Deliverables	Grand Average Tool Score	Summary of Tasks' Completion	Summary of Tasks Results vs. Delinquent Requests	Red	Yellow	Green	Corrective Action and Due Date Comments on Risk

The integrated system of tool, task, and gate deliverable scorecards provides a control plan for quantifying accrued risk in a traceable format that goes well beyond simple checklists. The task scorecard feeds summary data to this gate deliverable scorecard.

The next gate review format is used for executive reviews, where deliverables are summarized for rapid consumption by the executive gate-keeping team (see Figure 2.1).

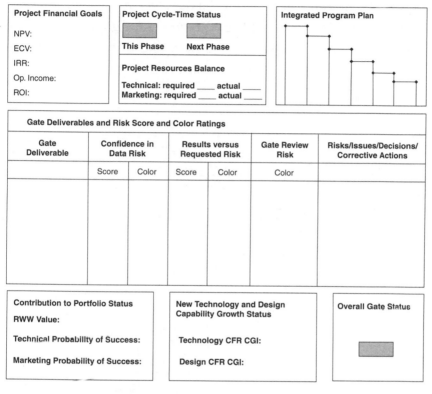

FIGURE 2.1   Strategic gate review scorecard.

Numerous companies committed to strategic portfolio management use this common format. The executive gate-keeping team looks at a number of these scorecards to balance risk across its portfolio of projects as the team seeks to grow the top line according to its business strategy. No project is immune from scrutiny or elimination if the summary data warrants that decisive action. Scorecards aid in decisiveness.

Summary

We have built an integrated system of scorecards (see Figure 2.2) that can be modified to suit any organization's phase-gate process. The design of scorecards must follow the design of your tool/task groups (clusters) and your deliverable/gate requirement groups as they are aligned with the phases of your product and technology portfolio-renewal process, product-commercialization process, and post-launch engineering process.

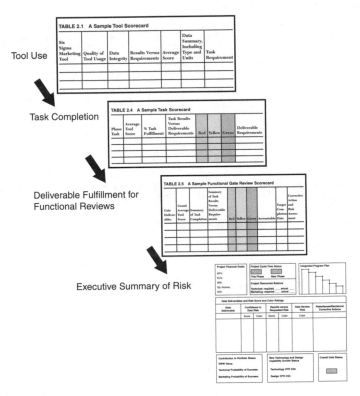

FIGURE 2.2   Integrated system of scorecards.

The system of scorecards can be used to manage the portfolio-renewal process. The first three scorecards are sufficient to manage risk across the portfolio-renewal process. As the portfolio-renewal team activates each technology-development and product-commercialization project, an executive summary scorecard is initiated and updated during technology development and commercialization.

As the old saying goes, "That which gets measured, gets done." We discourage the use of checklists and highly recommend the harder but more responsible use of a designed set of scorecards that will help you design what to measure and when to measure it so that you have a much higher probability of sustaining growth across your enterprise.

3

PROJECT MANAGEMENT IN TECHNICAL PROCESSES

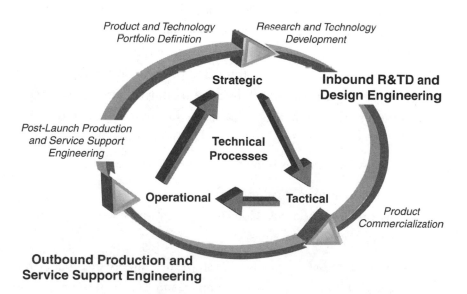

Designing cycle-time for strategic, tactical, and operational technical projects.

Six Sigma Contributions to Project Management in Technical Processes

Technical professionals have a great deal of standard material available to help them generate a project plan. It is not our intent in this chapter to review the basics of the project management body of knowledge. Our goal is to demonstrate how a few value-adding elements from traditional Six Sigma tools help in the design and analysis of technology development and product commercialization cycle-time. We want to be confident that what we choose to do and how long we forecast that it will take aligns with management expectations and market dynamics.

Over the last decade, while studying project cycle-time, we have found that executives and work teams are really out of touch with one another. This is true of both marketing and technical teams across the strategic, tactical, and operational processes. Executives have a cycle-time expectation that is almost always dramatically shorter than what their teams say they need to do the job right. Teams try to complete a set of tasks correctly and fully. Management forces these teams to do their work faster than that. Why does this happen? Usually because the business is doing too many projects and really does not know when each project rightfully must be finished and for what specific reasons. The thinking is, "The faster we get done, the sooner the cash will flow, and we have hope of making our numbers."

The results from this behavior are easy to define.

- Tasks that are critical go unfinished; a flow of work must stop before the tasks are done completely and correctly. Partial results are the outcome.

- A task that has seven essential steps to complete might get truncated. If you assessed the completion of each step, you would find that shortcuts were taken to get the work "done." Most senior executives have no idea how much of this "shortcutting" is going on inside their companies and should recognize how this is affecting their ability to meet growth goals. The work is *not* getting done correctly—it's done just to a point of "good enough for now—we will have to clean it up later when we have time (which is never because we have too many other projects to do so that we can make our growth numbers)." Execution means getting critical tasks done fully and correctly, not just fast. Growth is not sustainable under these self-defeating conditions. Stop rushing—just hurry according to a well-designed plan.

- Tools, methods, and best practices are used on an ad hoc basis. Poor alignment between tasks and enabling tools is the norm. Tools are not always viewed as important, and their intended contribution to creating proper results is underfulfilled.

- No systematic linkage of enabling tools is applied within the steps of completing a task or a series of related tasks. Tasks and tools are not aligned so that each part of a task gets done with the help of a proper and balanced set of tools. In fact, commonly critical tasks are improperly defined because they do not possess the discipline of a proper set of enabling tools that ensure that the task is developing the right results and summary forms of data. The tools that do get used are applied as a matter of personal preference instead of according to a standard of excellence that management universally recognizes. Each individual then must determine how to get tasks completed when it comes to supporting tools, methods, and best practices. Inconsistency is the standard mode of conducting work.

- Key deliverables are incomplete; at a gate review or milestone, the data executives or other leaders need to assess risk and make key decisions, which is simply not as helpful as it could be. This slows everything down—later.

- Summary data gets trumped up to appear to have enough credibility to enable the gatekeepers. Those responsible for gatekeeping get good at making incomplete data look like it is complete. We "save face." When presented with this situation, the typical gatekeeper immediately goes into a mode of using "best judgment." This is face-saving but is simply no substitute for balanced sets of data that really tell the truth. Using judgment to fill the gaps of incomplete sets of summary data is routine in modern corporations. When this becomes a standard practice of moving things along rapidly, consistently maintaining growth is not likely to occur. A mismatch arises between what data is required to make decisions and what data teams are allowed (funded and expected) to produce. Another problem is that data analysis lags decision points because we are too busy to get the analysis done; we decide without having facts as they really exist in the unanalyzed data!

We can do much to fix this mismatch. We need to discipline our management and work teams to *design* cycle-time, model it statistically, document critical path failure modes, and then negotiate what tasks and enabling toolsets are critical to producing the deliverables needed for proper decision making at a gate review or project milestone meeting. When this is funded and expected, risk is much easier to manage.

Designing Cycle-Time: Critical Paths of Key Technical Tasks/Toolsets by Phase

Designing cycle-time is different than simply creating a work breakdown structure and setting up a Gantt chart to show how long a series of tasks will take. Designed cycle-time has the following characteristics:

- **Gate requirements** are used to define very specific deliverables

- **Gate deliverables** are summary results that come directly from within-phase tasks that are done fully and completely. (What is worth doing is worth doing right.)

- **Critical tasks** are underwritten by a linked set of enabling tools, methods, and best practices that are proven to be effective in developing the data required for the gate deliverables.

- **Tool-task-deliverable sets** are integrated and linked for maximum value in producing gate deliverables and, consequently, fulfilling gate requirements (think "lean toolsets" in this context).

- **Balanced and trained resources** can be properly aligned with the detailed tool-task-deliverable sets (you know who and what you need to develop the gate deliverables—*core competencies!*).

Designing a critical path of cycle-time with these five characteristics greatly increases the likelihood of meeting gate requirements, which is a key step on the path to sustaining growth. This structured approach ensures two things:

- Real-winning-worthy projects from your activated portfolio of new or existing products (leveraged new line extension) can be fully supported for further investment for their contribution to fulfilling your growth requirements.

- Projects with too many critical gate deliverables that are falling short of meeting the gate requirements can be killed with confidence. You know that you are doing the right thing for the sake of the business.

"Designed" cycle-time buys you the assurance that a given project's true potential is being fully developed. With "undesigned" cycle-time, you end up pulling your punches and not giving your portfolio of projects the best chance of success. Diluted core competencies on ill-defined projects with anemic deliverables—how can that possibly sustain growth? We rush our way through too many projects for our available resources and hope we can make our growth numbers. When your "doers" see that their work life will be like this, they often just find another company to work for. That's yet another root cause for poor sustainability in growth: You can't retain the brightest and the best.

Nine Key Steps

"Designing" cycle-time correctly involves nine key steps. We have highlighted four specific value-adding elements out of the nine that we feel are derivative from Six Sigma methods. This list takes real discipline to put into practice. Our clients who have done these nine items are successfully growing; those who have thrown in the towel are not.

1. Provide a clear and complete definition of gate requirements.

   What exactly do you need to get through the gate? (What do you need to know to manage risk and make key decisions?)

2. Provide a clear and complete definition of gate deliverables.

   These include documents that contain summary data sets representing the progress against gate requirements (the truth lies in this data, not in your opinion about data that does not exist).

3. **Create detailed process maps, work break-down structures, and designed workflow charts** (very Six Sigma–ish items to show how a technical process has controllable and uncontrollable inputs and outputs).

   This concerns the flow of critical technical tasks within each phase (a system of integrated, value-added work for a phase).

4. **Link major tasks to balanced sets of tools, methods, and best practices** (a value-adding approach similar to Lean DMAIC Six Sigma projects).

This involves enabling toolsets that ensure that the right data will be fully developed so that a task is truly finished (this is how tasks really produce data-based deliverables in your phase-gate process).

5. Generate a RACI matrix to define personal accountability for task/tool completion.

A Gantt chart of tasks aligned with who specifically is **responsible** for doing the work, who is **accountable** for the resources doing the work, who will act on a **consultory** basis to help get the work done, and who must be **informed** of task progress and results in parallel or dependently related tasks (a what/who matrix so everyone knows what they are accountable for, to avoid miscommunication and failure to execute at a detailed level of task completion).

6. Generate a PERT chart for the phase.

This "network" diagram illustrates the serial and parallel flows of tasks as they really occur in sequential time (a visual flow of work to show task relationships, dependencies, and buffers).

7. Calculate the critical path for the phase.

This is a timeline of critical tasks that is the longest between starting and ending a phase (the tasks that define the cycle-time for a phase).

8. **Conduct Monte Carlo simulation of the cycle-time distribution forecast for the phase** (a value-adding statistical analysis and forecasting $[\Delta Y = f(\Delta X)]$ tool common in Six Sigma applications).

This frequency distribution of likely times characterizes the probability of finishing the critical path of tasks (a range of times over which we have quantifiable confidence in being done with our critical tasks).

9. **Create a project cycle-time failure modes and effects analysis for the tasks on the critical path.** (FMEA is a very common tool used in every Six Sigma project known to mankind.)

This involves a detailed assessment of what can go wrong for each task on the critical path, a prevention plan to help ensure that tasks stay on track as much as possible, and a reaction plan for when they don't (a risk-mitigation plan).

When these nine items are done well, teams have much higher credibility when they commit to a project schedule and the time it is *likely* to take to complete it. Start eliminating each of the nine one-by-one, and see how comfortable you feel about meeting your commitment to what you said you would do. Gone are the days of deterministic (single point in time) time projections: Face into the truth and enjoy the credibility and confidence you have when you look at the facts about how long things really take to complete correctly. Don't rush your people.

Modeling Technical Task Cycle-Time Using Monte Carlo Simulations

As we said earlier, four of the nine items go beyond traditional project-management methods to really add value in producing high-integrity project cycle-time planning and results. Conducting a series of Monte Carlo simulations on the designed critical path of tasks is one of them and is a big deal. Here we go ....

A Monte Carlo simulation is a method for changing (varying) the time allocated for each task (ΔX) on the critical path using carefully selected limiting conditions and then adding up the entire time for the phase to be completed (ΔY). The value-adding elements of a Monte Carlo simulation on designing cycle-time are summarized as follows:

1. A math model that adds up each task duration for the critical path is defined.

   $\Delta Y = f(\Delta X)$, where ΔY is total cycle-time range and each ΔX is a task duration range on the critical path

   $$\Delta Y = \Delta X_{task\ 1} + \Delta X_{task\ 2} \cdots + \Delta X_{task\ n}$$

2. Each task is represented as a triangular distribution of possible durations, typically in units of days or hours (ΔX values).

   Possible task durations are statistically represented using triangular distribution as illustrated below in Figure 3.1 to represent a linearly diminishing likelihood of a task duration occurring at the ends of the triangle. The "peak" of the triangle represents the most likely time a task will take to complete.

Example of Crystal Ball MC Software from Decisioneering

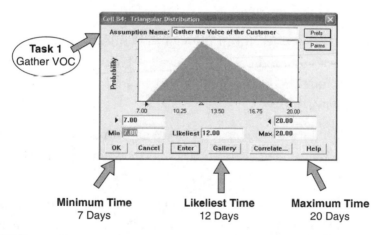

FIGURE 3.1   Monte Carlo simulation assumption entry window.

3. Each triangular distribution is "loaded" with the best estimates of three types of durations that can be estimated for each task.

   The estimates must come from a team of veterans of the technical process that have experience in how long tasks and tools take to complete. Garbage in equals garbage out—of course, we prefer to work with $\Delta Y = f(\Delta X)$. Know your tools and how long they take to enable the completion of tasks.

Time is specified for a triangular distribution in three ways:

- Shortest likely time to complete a task (when things go really well—yeah, right!)

- Most likely (mean or median) time to complete a task

- Longest likely time to complete a task (when things go really wrong—Murphy's Law awaits!)

4. The Monte Carlo simulation samples or selects at random a duration from within the triangular distribution representing each task's possible range of durations.

   The software running the simulation uses a random number generator to "select" an unbiased duration from within each triangular distribution.

5. The sum of each run through the critical path of randomly selected task durations is calculated and entered into a histogram (frequency distribution chart for ΔY, as shown in Figure 3.2).

6. The final frequency distribution estimates a range and frequency of distributed cycle-time end points for the phase (ΔY).

   The shape (usually a reasonably normal distribution) and its statistical parameters are graphically presented for analysis (enlightened discussions about a range estimate of reality).

7. The percent confidence in finishing the project on or before a specific date can be viewed on the frequency distribution chart.

   If you don't like the answer, you can redesign the critical path until you reach a compromise on what to do, when to do it, and how long it must take on a task-by-task basis.

$$\Delta Y = \Delta X_{task\ 1} + \Delta X_{task\ 2} \cdots + \Delta X_{task\ n}$$

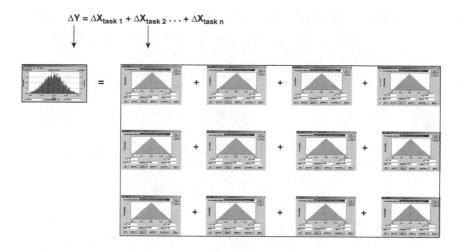

Example of Crystal Ball MC Software from Decisioneering

FIGURE 3.2 Monte Carlo simulation model.

You can shift down the mean or median of the cycle-time frequency distribution (Y) by changing the mean or median duration values (most likely Xs).

You can narrow the width of the forecasted distribution by changing the values of the shortest and longest likely times for each task in the critical path (as shown in Figure 3.3).

8. Use the project FMEA to help rationalize changes made to the critical tool-enabled task sequences that give you the final cycle-time risk you can bear (yes, it will be longer than the old way, but it's a lot more credible).

   • Decide what tools, methods, and best practices are "must haves" vs. ones that are optional (lean out the cycle-time).

   • Decide what tasks are absolutely critical to be done fully and completely vs. those that can be done with lesser levels of completeness (more leaning out the cycle-time).

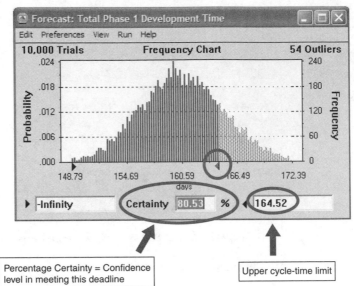

Example of Crystal Ball MC Software from Decisioneering

FIGURE 3.3   Monte Carlo simulation certainty output.

If this sounds a lot like "Lean," you are correct; this is Lean Six Sigma when applied to project management. When this kind of information is available for cycle-time discussions for teams, their project managers, and executives, a much better dialog arises. Realistic negotiations and enlightened trade-offs can take place. If you can't see how you could possibly spend the time to do such diligence in cycle-time design and management, we wish you good luck—you will need a lot of it because luck is now your formal strategy for sustaining growth. We hope you choose to enhance your luck with these nine opportunities to prevent downstream cycle-time problems.

Documenting Failure Modes in the Critical Paths of Technical Tasks by Phase

The last unique value-adding method of traditional project management is the application of failure modes and effects analysis to the design of cycle-time. This one item almost never arises in project management, yet it is the easiest to apply and offers a really good payback for the time you invest in doing it.

Applicable to any process, FMEA helps identify risks and their consequences. The key enhancement we like about it is that it helps teams design a proactive approach to *avoiding* cycle-time problems. Yes, it can help devise reaction plans to work around problems when they arise, but that is not the best way to use the project FMEA. A typical format for a project FMEA is illustrated in Table 3.1.

A project FMEA has six basic elements and three quantifiable attributes to identify just how big your risks are.

1. What is the task (work function) that can fail?

2. What is the exact nature of the failure mode for a task?

3. What are the specific effects that come after the failure has occurred?

4. What level of risk does the failure present to the team?

5. What preventative and reactive control can you put in place to lower the level of risk?

6. What does a risk-reduced project plan look like at a gate review?

- Every team should present this plan at the gate review just before the next phase of work.

- Gatekeepers should offer help and resources to ensure that cycle-time problems are prevented or avoided for the next phase.

- A good project leadership team is always at least one phase ahead of the working team, clearing obstacles, providing all that is needed for project team success, and killing low-value-add projects that sap strength from strategic-growth projects. (We know this is hard to do.)

TABLE 3.1   Project FMEA Format

Task/Tool	Tool Function	Potential Failure Mode	Potential Failure Effects	SEV	Potential Causes	OCC	Current Project Management Evaluation or Control Mechanism	DET	RPN
What is the Task/Tool application under evaluation?	What is the purpose of the tool?	In what ways does this Task/Tool application corrupt cycle-time goals?	What is the impact on the project cycle time?	How severe is the effect on the schedule?	What causes the loss of Task/Tool function?	How often does cause or FM occur?	What are the tests, methods, or techniques to discover the cause before the next phase begins?	How well can you estimate cause?	
									0
									0
									0
									0
									0
									0

During the analysis stage of FMEA, risk is quantified in three ways:

1. The severity of the failure mode

2. The frequency of occurrence of the failure mode

3. The detectability of the failure

   • Can you see it coming? Measure impending failures.

   • How well can you measure the failure after it has occurred?

Each of these items is typically laid out in a table of decreasing magnitudes using a scale of 1 to 10. A rating of 10 indicates high risk; 1 indicates low risk. The three values are multiplied to yield a risk priority number (RPN). A high RPN is a call to action to devise a preventative and reactive plan to mitigate the risk. RPNs are used to rank and prioritize the critical path of tasks based upon contribution to risk. Every task on the critical path should have an RPN and a control plan. This should be a non-negotiable requirement for every gate review. The summary gate deliverable is the cycle-time FMEA ranks and the control plans (especially the "prevent" plan). The gatekeepers can go down the critical path by ranked risk levels and decide what they can specifically do to help the team protect its forecasted cycle-time. This helps the team stay unified, get predictability designed into the work, and do everything possible to stay on time.

Integrated Project Plan

An entire project of standard technical work (Figure 3.4) can be designed, forecasted, modified, and balanced for the right cycle-time to meet the goals of the project.

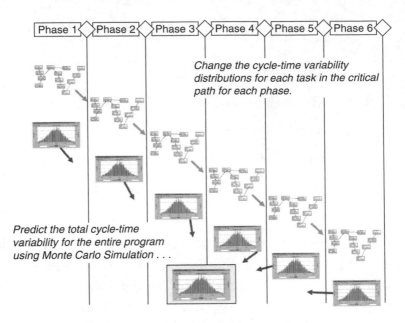

Example of Crystal Ball MC Software from Decisioneering

FIGURE 3.4   Integrated project cycle-time model.

With this kind of disciplined approach to designed cycle-time, executives, project managers, and the actual team of technical professionals can fully communicate just how long a project should take from a realistic position of tool-task flow to produce the right deliverables in fulfillment of the gate requirements.

4

STRATEGIC PRODUCT AND TECHNOLOGY PORTFOLIO RENEWAL PROCESS

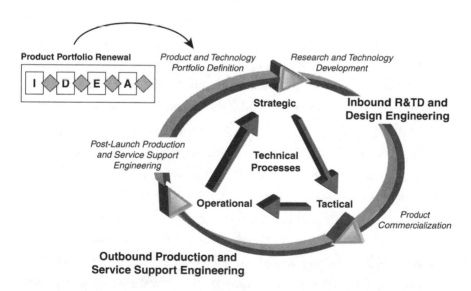

Inbound technical contributions for product and technology portfolio definition and renewal.

Six Sigma in the Strategic Product and Technology Portfolio Renewal Process

We call this part of the work that technologists do "inbound." Inbound work means technologists go out, gather data, and bring it back to the company to help define, develop, and renew the product portfolio and its supporting technology portfolio. Who do they go out with? They travel, work, and consolidate data with their peers from inbound marketing and product planning as an integrated team. Technologists, product managers, market research specialists, attorneys, sales managers, production and supply chain specialists—those who have value to add—participate in this most strategic of data-gathering and processing processes. This work takes money, time, and really top-notch people. Underinvesting here is the beginning of an ingrained inability to sustain growth. Ask yourself right here at the outset, "Am I understaffing and underinvesting in this strategic work?"

Technology portfolios have no meaning outside the product portfolio they enable. Product portfolio requirements primarily drive technology portfolio requirements. Can a new, innovative technology drive product portfolio requirements? Yes. How often is this the case? Not as often as you might think, but it does happen. Therefore, we must have an open mind and balance our behavior accordingly.

Inbound is how we characterize the flow of data that will be acted upon to define the families of new products and services that a company will convert into financial growth by way of satisfied customers. Companies don't exist to satisfy technologists who develop new technologies. If you are a technologist or engineer and you don't get a kick out of seeing your customers delight in what you develop, you need to reconsider why and where you are working. In industry, you have to care deeply about your customers and invest heavily in the data that guides what ideas you care about and work on. Because the data matters, so does the value-adding tools from Six Sigma.

This inbound work intentionally creates data that helps the team develop desired forms of variation. Here, Six Sigma helps to identify forms of variation that actually create value. Product portfolios are, by nature, a variety of offerings based upon a blend of technology dynamics and customer behavior dynamics. Sometimes technology drives customer dynamics, and sometimes customer dynamics drive technology development. Rarely is technology 100 percent dominant as a driver of customer behavior. Customer behavior generally takes the lead in the dance of sustained growth.

This strategic set of tasks requires a close partnership with technologists and marketing and product-planning professionals. Strategic inbound marketers focus on characterizing emerging customer behavioral dynamics, competitive marketing trends, and brand dynamics. Strategic, inbound technologists focus on characterizing emerging technology dynamics, competitive technology trends, and intellectual property dynamics.

Technology is an enabler. It is developed and transferred. Sometimes it is sold or licensed before being commercialized—that helps enable growth in some companies. The aggregate variety of technologies that are developed is called the technology portfolio. R&D can create it from scratch with little influence from the market or business leadership, or by orders from the product portfolio renewal team based upon requirements they develop and give to R&D. If your technology portfolio is developed in isolation from your product portfolio, there is less hope you can sustain growth. A balance must be achieved between technology pushed from within R&D and technology pulled from the product portfolio renewal team. We know that both are good and necessary. This chapter illustrates how Six Sigma methods can help technologists develop the data needed to do the following:

- Identify opportunities for "pushed or pulled" technology to impact growth

- Define technology requirements and activation of technology development or acquisition projects

- Evaluate product portfolio architectural alternatives in light of the developing technology portfolio

- Activate product commercialization projects based upon the maturity of the technology development projects

Product Portfolio Renewal

These four phases of work form the basis of how data can be developed in reasonable batches so that the strategic portfolio renewal team can manage risk and make decisions. Take note: Traditional DMAIC Six Sigma has no meaning here. Six Sigma discipline and methodologies can help this team do the right things at the right times so that its data is high in integrity and low in vulnerability. Nobody naturally defines the process of portfolio development and renewal by the DMAIC steps; the work flow and nature differ from the problem-solving steps of DMAIC. We have great respect for the DMAIC process, but this is not its natural application environment, so don't force its steps to define the work flow conducted across the portfolio-renewal process.

This chapter focuses on the arena of strategic product and technology planning that we call product and technology portfolio renewal (P&TPR). P&TPR is conducted using a strategic inbound, hybrid, technical and marketing process we define using four distinct phases called IDEA (Identify, Define, Evaluate, and Activate). We fully define these phases shortly. If you don't like these names for the phases of portfolio renewal, use ones you like. What matters is that you name your process phases after the major things you do as you progress though the clusters of work that define how you assess risk and make key decisions across the entire process.

Growth has three process arenas that contribute to its sustainability: strategic, tactical, and operational. They must all be linked by the flow of data and a system of integrated metrics (leading indicators) to create and track consistent progress that ultimately will produce designed gains in growth. If you let any one of the three languish in performance, growth will be come erratic—perhaps even out of control. Think of this strategic, front-end process as the beginning quarter of a control plan for growth.

Measuring results from the work that technical, marketing, and other professionals do depends upon the linkage and flow of data within and across these processes. Predictable results come from a well-planned series of tasks. Tasks conducted by technologists and engineers must be measured and controlled as a consistent and integrated system as they flow from strategic to tactical to operational processes.

Process Discipline in Product and Technology Portfolio Renewal

Tackling the job of adding process discipline and metrics to the tasks conducted within a product and technology portfolio renewal process is by far the most difficult task when integrating Six Sigma into the four processes we address in this book. Adding process metrics and discipline to R&D functions in any environment is not a trivial matter. It must be done carefully, taking into account the unique, creative culture that exists in these types of organizations. Art and science both are involved here. Attention must be paid to how technical experts interact with other teams within adjacent business functions, especially marketing professionals working in tandem with them in portfolio planning. Cookie-cutter approaches will not work. The process must be custom built. The alignment of Six Sigma tools and methods must be designed into the custom flow of work.

Many companies have a measurable lack of process rigor in their product and technology portfolio renewal process. It is a tough environment to get alignment across organizational boundaries. This is especially true of inbound marketing and R&D. A lot is at stake when structuring the future of the company in the form of its product and technology portfolio architectures. If the two architectures are incongruent, it is very hard to sustain growth. Oddly enough, the place where the largest risk lies for developing strategic growth is the very place where process discipline and organizational collaboration contains the least linkage, rigor, staffing, and metrics. The most notable outcome is that companies take on too many projects in a hope that enough of them will deliver financial results that will meet the growth requirements from the business strategy. This mainly happens because of the emotion we know as fear. Management overloads the workforce, particularly in understaffed marketing organizations.

Another significant issue is the cycle-time disruption from technology readiness and transfer being out of synch with the activation of product commercialization projects. Typically the technology needed to enable a new product gets developed alongside the new product; risk is high and hope is the emotion in use. We call this approach synchronous technology development. In some cases, this is a sensible risk to take, such as when the new technology has low to moderate risk that it won't fulfill the product's requirements. Asynchronous technology development should be conducted to minimize risk when the new technology contains signs of moderate to high risk. How do you tell the difference? You have to invest in data.

Six Sigma tools can smoke out technology sensitivities to quantify just how risky a new technology really is. Newly transferred technology is frequently immature and debilitates the product delivery schedule because the technology has to be redone. Executives expect outstanding results from project teams. Commercialization teams can't possibly complete all their tasks that should be done to underwrite these expectations when nonrobust, nontunable technology is prematurely integrated into a commercialization project. Executives want to see orderly activation, design, and launch of the product lines. If the product portfolio and technology needed to enable it are not linked and timed for integration, the work of executing the new portfolio just cannot happen correctly, let alone happen on time. We need to design strong, strategic alignment between product and technology portfolio architecting tasks for the sake of downstream cycle-time efficiency and control. Too many projects, some with technology time bombs ticking away inside of them, lead to a natural outcome: unpredictable execution and growth.

The product and technology portfolio renewal process is the first of two strategic processes in which R&D professionals can use Six Sigma methods. The second process is the formal development of new technologies that the product and technology portfolio process requires (see Figure 4.1).

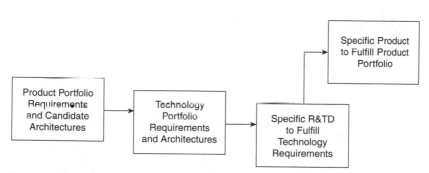

FIGURE 4.1   Strategic to tactical workflow.

When we define inbound technical processes, research and technology development is the strategic component, while product design engineering, done during commercialization, is the tactical component.

Figure 4.2 illustrates the inbound macro tasks that define what must be done during product (white boxes) and technology (gray boxes) portfolio definition and development.

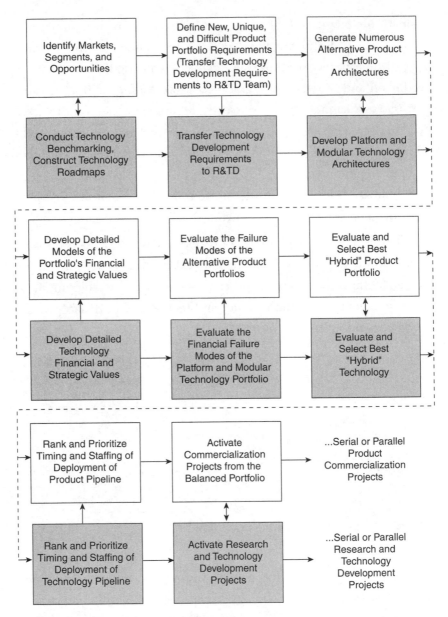

FIGURE 4.2   Integrated marketing and technical workflow.

Stepping back to develop more context, Figure 4.3 shows the big picture of integrated marketing and technical functions that reside within the inbound and outbound technical arenas.

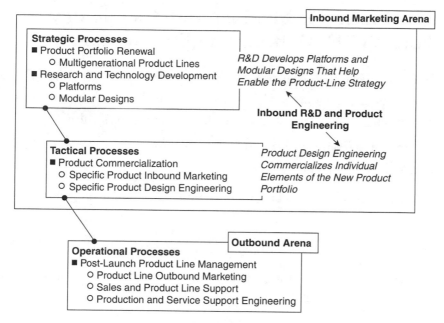

FIGURE 4.3   Process linkage diagram.

Marketing and technical processes and functions must be linked for Six Sigma in marketing, R&TD, and design to enable growth. Integrated, multifunctional teams from inbound marketing, R&TD, design, and production/service support engineering must be used across all three process arenas to develop and share data to manage risk and make decisions. You must recognize that the processes to sustain growth are far more complex and require broader toolsets than the simple DMAIC problem-solving steps can handle. DMAIC can solve process problems that you encounter along the way, but it is not *the* way. We need properly-designed phases and gates to manage risk and make key decisions across these process arenas. The form of Six Sigma we are discussing aligns tools to tasks that we conduct on a day-to-day basis within a designed phase-gate process. To design a process like the one we are about to discuss, you can use the well-known DMADV approach (as suggested in Chapter 1, "Introduction to Six Sigma for Technical Processes").

The Phases of Product and Technology Portfolio Definition and Development

In the *strategic* technical process environment, we have identified a generic process segmented into four distinct phases (these are suggested names—use what you prefer in your company's culture).

The IDEA process for product and technology portfolio definition and development:

1. **Identify**—Markets, their segments, and the opportunities they contain

   - Technology benchmarking and roadmapping

2. **Define**—Portfolio requirements and product portfolio architectural alternatives

   - Technology requirements transfer, platform and modular technology architecting

3. **Evaluate**—Product portfolio alternatives against competitive portfolios

   - Platform and modular technology portfolio evaluation and selection

4. **Activate**—Ranked and resourced individual product commercialization projects

   - Technology project ranking, staffing, and activation

A *phase* is a period of time that is designed to conduct work to produce specific results that meet the requirements for a given project. This book takes the view that every cycle of strategic portfolio renewal is a project with distinct phases and gates. A *gate* is a stopping point to review results against requirements for a bounded set of portfolio-renewal tasks. A phase is normally designed to limit the amount of risk that can accrue before a gate-keeping team assesses the summary data that characterizes the risk of going forward. Six Sigma enhancement helps make gatekeepers more decisive because of the integrity of the data they consume. The portfolio renewal and product commercialization processes are kept "under control" through a system of phases and gates that define the control plan for the inbound technical and marketing processes.

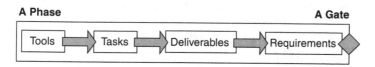

FIGURE 4.4   Tools, tasks, deliverables linked to requirements.

Phases are best described by how they integrate tools, tasks, and deliverables to meet gate requirements. Gate-keeping teams establish gate requirements and the deliverables that fulfill them, which, in turn, define the context for exactly what tools and tasks are appropriate for the design of the project cycle-time. These tool-task flows are designed to define the critical parameters (deliverables) for the inbound portfolio team.

The first phase in the IDEA process is heavily enabled by inbound marketing tasks and tools. It is also enabled by inbound technology roadmapping and benchmarking tasks and tools.

Before entering the IDENTIFY phase, the business strategy must be clearly defined. This breaks down into several general areas:

- Financial growth goals

- Core competencies and capabilities

- Innovation strategy (including both marketing and technology)

Strategic inbound technical and marketing contributions for portfolio renewal will be rudderless without these elements clearly documented.

The IDENTIFY phase includes three main requirements:

1. Identify the general market or markets in which you intend to invest and win.

   - Markets that contain profit and growth potential

2. Identify the specific segments within the general markets.

   - Segments that contain differentiated needs and dynamics

3. Identify specific opportunities within and across the segments.

   - Opportunities defined as the market's financial potential linked to needs that are unfulfilled, as well as behavioral dynamic trends linked to trends in technology dynamics and emergent technology break-throughs

Opportunities can further be broken down into external and internal forms:

- Financial and need opportunities that the markets and segments possess in relation to your competition, defined by market dynamics (these must align with and sum up to fulfill the business's financial growth goals)

- Technology and service opportunities that you are capable of delivering, defined by your ideas, as embodied in your management and control of critical parameters for technology and service dynamics (these must align with the innovation strategy and the strategic core competency the company has)

The second category of opportunities can be expressed as "ideas" that are gathered, documented, and converted into product and product line concepts that will be enabled by new technology platforms and modular designs (see Figure 4.5). The first category of opportunities can be converted into requirements that, in the aggregate, will form the new product portfolio requirements. With ideas and product portfolio requirements in hand, you can conduct product portfolio architecting, a major goal of the DEFINE phase for the IDEA process.

| Opportunities are converted into requirements | ⟹ | Ideas are converted into offerings concepts |

FIGURE 4.5   Conversion diagram.

P&TPR is concerned with identifying, defining, and evaluating external opportunities linked to your internal ideas. At 3M, Six Sigma is used strategically to double the number of these linked opportunities and ideas before commercialization. Tactically, 3M attempts to launch products that now have three times the return compared to their previous launches that did not have the enablement of Six Sigma.

The DEFINE phase is the key transfer point for delivering product portfolio requirements to the R&TD organization. R&TD receives these diverse requirements and translates them into technology requirements. This key requirement transfer and translation process helps define the technology portfolio. The integrated team reviews the translation to ensure that the basis for developing and renewing the technology portfolio aligns with the guiding requirements, as shown in Figure 4.6, that will be used to define numerous architectural alternatives for the product portfolio.

With several alternative product portfolio architectures defined, the team enters the EVALUATE phase. This phase involves the data-driven evaluation of the candidate portfolio architectures against competitive benchmarks in light of the portfolio requirements. A superior hybrid portfolio architecture emerges from this process phase.

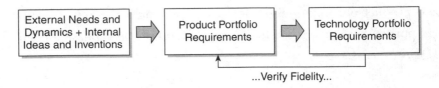

FIGURE 4.6 Needs/ideas to requirements.

The final phase of P&TPR is to ACTIVATE product-commercialization projects out of the superior portfolio architecture. The focus here is on activating projects that will, in the first phase of commercialization, convert opportunities into *specific product* requirements and ideas into *specific product* concepts. Frequently, one hears of a specific product being done as a DFSS (design for Six Sigma) project. DFSS product-commercialization projects are activated from the ranked and prioritized portfolio of projects from P&TPR. It is important to note at this point that portfolio requirements are different than specific product requirements. Portfolio requirements are the focus of P&TPR, while product requirements are the work of tactical commercialization teams after a specific product commercialization project has been activated from the portfolio.

The IDENTIFY Phase of Product Portfolio Definition and Development

Product Portfolio

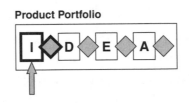

Producing the IDENTIFY phase deliverables requires an investment in numerous detailed, inbound marketing and technical tasks enabled by specific tools, methods, and best practices. Their integration ensures that the right data is being developed and summarized to fulfill all three major requirements at the IDENTIFY gate review.

The three major requirements for the IDENTIFY gate include the following:

1. Identify the general market or markets in which you intend to invest and win.

   • Markets that contain profit and growth potential

2. Identify the specific segments within the general markets.

   • Segments that contain differentiated needs

3. Identify specific opportunities within and across the segments.

   • Opportunities defined by segments that contain or bound distinct pools of financial potential linked to needs that are unfulfilled

The deliverables at the IDENTIFY gate include the following:

Major deliverables:

• Documented growth goals (financial targets and tolerances)

• Documented core competencies (availability and readiness)

• Documented innovation strategy (marketing and technical components)

• Documented markets (macro profit pools)

• Documented market segments (micro profit pools)

• Documented opportunities (financial potential, customer need, and technology dynamics)

Supporting IDENTIFY gate deliverables:

• Market and segment behavioral dynamics map

  • Customer behavioral process maps

  • Value chain diagrams

  • Key events timelines

- Technology dynamics map
  - Technology roadmaps
  - Technology value chain diagrams
  - Key events timelines
  - IP trends and patent dynamics/trends
- Competitive benchmarking data and trend analysis
- Market perceived quality profile and gap analysis
- Porter's Five Forces analysis and segmented risk profiles
- SWOT analysis and segmented risk profiles

The tasks to be performed within the IDENTIFY phase that will produce the deliverables include these:

1. Define business strategy and financial goals.
2. Define innovation strategy.
3. Define markets.
4. Define market segments.
5. Define opportunities across markets and within segments.
6. Gather and translate the "over-the-horizon" voice of the customer data.
7. Document new, unique, and difficult customer needs across segments.
8. Construct market and segment behavioral dynamics maps.
9. Construct technology dynamics maps.
10. Conduct competitive benchmarking (marketing, sales, and technical).
11. Create database of internal ideas based upon opportunity categories.
12. Create DEFINE phase project plan and risk analysis.

The tools, methods, and best practices that enable the tasks associated with fulfilling the IDENTIFY deliverables include these:

- Market identification and segmentation analysis
  - Secondary market research and data-gathering tools
  - Economic and market trend forecasting tools
  - Statistical (multivariate) data analysis
    - Cluster, factor, and discriminate analysis
- VOC gathering methods
  - Primary market research and data-gathering tools
- Market and segment VOC data-processing best practices
  - KJ diagramming method
  - Questionnaire and survey design methods
- SWOT analysis method
- Market perceived quality profile method
- Porter's Five Forces analysis method
- Market behavioral dynamics map methods
- Technology dynamics map methods
- Competitive benchmarking tools
- Idea capture and database-development tools
- Project-management tools
  - Monte Carlo simulation (statistical cycle-time design and forecasting)
  - Failure modes and effects analysis (cycle-time risk assessment)

Control, management of risk, and key decision making are greatly enhanced when we clearly define the requirements, deliverables, tasks, and enabling tools within the IDENTIFY phase of P&TPR (see

Figure 4.7). If these four layers are ambiguous, detailed root causes of poor sustainability in growth begin to emerge. When these four elements begin to degrade, you have begun the slippery slope down the road to poor performance that will eventually result in erratic growth.

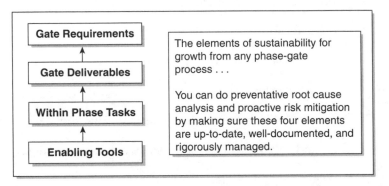

FIGURE 4.7   Tool, task, deliverables, requirements flow diagram.

Six Sigma discipline is commonly associated with the design and application of a control plan as a process is established, stable, and known to be capable of meeting requirements. This is a control plan for conducting enhanced strategic marketing and technical functions in the portfolio-renewal process. We often hear of control plans for manufacturing processes. This is an equivalent control plan for conducting work in a phase-gate process environment. It applies across strategic, tactical, and operational processes.

Table 4.1 integrates the Identify phase requirements, deliverables, tasks, and enabling tools.

TABLE 4.1   Tool, Task, Deliverables–to–Requirements Integration Table

Requirement	Deliverable(s)	Task(s)	Tools, Methods, and Best Practices
Identify the general market or markets.	Documented markets (macro profit pools)	Define markets.	Market identification analysis Secondary market research and data-gathering methods
	Documented revenue growth and profit potential	Define growth and profit potential.	Economic and market trend forecasting tool
	Documented market behavioral dynamics map • Customer behavioral process maps • Value chain diagrams • Key events timelines	Construct market behavioral dynamics map. • Gather and translate VOC data. • Construct value chain diagrams. • Construct key events timelines.	Process mapping Customer ID matrix Customer interview guides Customer interviewing methods KJ analysis Questionnaire and survey design and analysis
Identify the specific segments.	Documented market segments (micro profit pools)	Define segments within broad market(s).	Primary market research and data-gathering methods Statistical (multivariate) data analysis Cluster, factor, and discriminant analysis

continues

TABLE 4.1 Tool, Task, Deliverables–to–Requirements Integration Table (continued)

Requirement	Deliverable(s)	Task(s)	Tools, Methods, and Best Practices
	Documented needs that are different between segments	Define new, unique, and difficult needs within each segment.	KJ analysis NUD need screening KANO analysis
	Documented segment behavioral dynamics map • Customer behavioral process maps • Value chain diagrams • Key events timelines	Construct market behavioral dynamics map. • Gather and translate VOC data. • Construct value chain diagrams. • Construct key events timelines.	Process mapping Customer ID matrix Customer interview guides Customer interviewing methods KJ analysis Questionnaire and survey design and analysis
Identify specific opportunities.	Documented opportunities (financial potential, unfulfilled customer needs, and competitive and internal technology dynamics)	Summarize all opportunities.	
	Documented technology dynamics map • Technology roadmap • Technology value chain diagrams • Key events timelines • IP and patent trends	Construct technology dynamics maps.	Process mapping Technology roadmapping Value chain diagramming Key events analysis Patent searches and analysis

	Documented competitive benchmarking data and trend analysis	Conduct competitive benchmarking (marketing, sales, and technical forms).	Competitive benchmarking methods
	Documented market perceived quality profiles and competitive gap analysis	Conduct MPQP.	Market perceived quality profiling
	Documented Porter's Five Forces analysis and segmented risk profiles	Conduct Porter's Five Forces analysis.	Porter's Five Forces analysis method
	Documented SWOT analysis and segmented risk profiles	Conduct SWOT analysis.	SWOT analysis method
	Documented internal database of product ideas	Conduct idea generation.	Idea capture and database-development tools TRIZ, brainstorming, etc.
Define Phase Project Plan	Documented Define Phase Project Plan	Conduct Define phase project planning.	Project-management methods Monte Carlo simulation Critical path/chain FMEA

If your people don't know how to conduct these tasks with the help of these tools to produce the deliverables, you lack functional excellence at the tool-task-deliverable level. If your people can technically do these things but you don't require the deliverables, you lack management excellence. Management excellence is all about developing and demanding the fulfillment of clear requirements. Functional excellence is all about developing a core competence of generating deliverables that fulfill the requirements based upon disciplined tool-task performance under realistically designed project timelines.

The DEFINE Phase Tools, Tasks, Deliverables, and Requirements

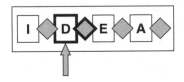

Producing the DEFINE phase deliverables requires an investment in numerous detailed inbound marketing and technical tasks enabled by specific tools, methods, and best practices. Their integration ensures that the right data is being developed and summarized to fulfill the major requirements at the DEFINE gate review (see Figure 4.8).

The major requirements for the DEFINE gate include the following:

1. Define the new, unique, and difficult (NUD) product portfolio requirements.

    - New requirements (targets and ranges) that no one in the competitive or adjacent industries are fulfilling today. They often call for invention or significant innovation in new technologies.

    - Unique requirements (targets and ranges) that your competitors or adjacent industries are fulfilling today but you are not. Usually these require technology licensing or innovative leveraging of existing technologies.

- Difficult requirements that are specific opportunities for you if you can use your core competencies to overcome these hurdles. Many technology requirements are found in this category as hybrid new and difficult requirements (most things that are new are very difficult).

2. Define candidate product portfolio architectures (various mixes of ideas).

- The variety of opportunities are matched with your various ideas.

- Several different product portfolio architectures are generated to fulfill the portfolio requirements across and within the markets and their segments.

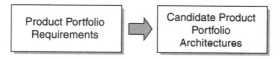

FIGURE 4.8   Proper flow for identifying candidate portfolio architectures.

The flow of knowledge into the DEFINE phase includes the following elements in Figure 4.9.

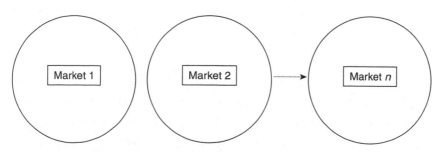

FIGURE 4.9   Market identification.

First, we have defined markets where there exists sufficient profit potential to fulfill our business strategy. These markets possess new, unique, and difficult needs that align with our core competencies.

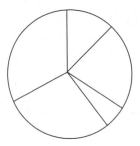

FIGURE 4.10   Market segmentation.

Next, we have determined, with statistically significant data, that we have defined differentiated needs between the market segments within the general markets. We have diverse NUD needs ready to translate into a set of product portfolio requirements that possess common and differentiated targets and fulfillment ranges. These requirements drive the product portfolio-architecting process. Innovation at this level is the most strategic form of creativity and ideation a company can conduct. If you fail to innovate here, the sustainability of growth is at risk.

Matching ideas to opportunities within and across segments helps ensure that you are generating the right mix of alternatives for your portfolio-architecting process. Blending and considering alternatives between your ideas and market opportunities helps lower the risk of launching an unbalanced portfolio. This helps stabilize growth potential on a systematic basis.

Phase work integrates and documents mixes of ideas that are aligned with the real opportunities that exist within and across your markets and segments. Innovative ideas are key to exploiting market opportunities. Constant creation, gathering, prioritizing, and pruning of ideas is part of the dynamics that make companies grow. If there is a lull in this activity, a dip in growth can occur. If you let your ideation die out for a period of time, you will see a direct cause-and-effect relationship relative to erratic growth. Ideas can come from anywhere, and constant incubation transcends the IDEA phases. Harvesting and integrating ideas into candidate portfolio architectures in the DEFINE phase is a must for the IDEA process to work properly.

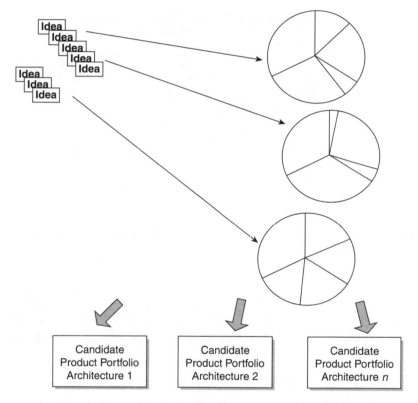

FIGURE 4.11   Idea matching to segments and architectures.

To summarize at this point, you can see how needs drive ideation. Ideas can be linked and matched to opportunities for portfolio architecting. Portfolio requirements and alternative architectures drive technology portfolio planning and development.

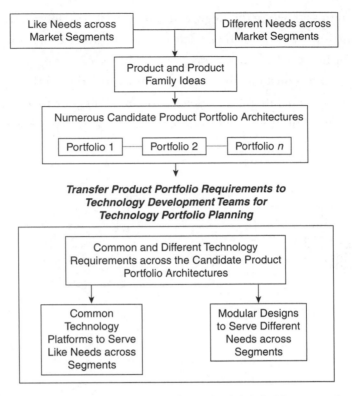

FIGURE 4.12 Transition from product portfolio definition to technology portfolio definition.

The P&TPR team defines the requirements for the new product portfolio architecture, based upon NUD needs from the segments. Numerous ideas are matched with market opportunities that bound the target environment for the new portfolio. The ideas are then grouped and linked to form candidate product portfolio architectures. Each architecture represents one mix of product ideas that has the potential to fulfill the business strategy and meet the aggregate financial growth goals. When a number of candidate portfolio architectures are defined, they and the portfolio requirements are transferred to the teams that conduct technology portfolio planning and development. The early work of technology development is now linked to the output from the IDEA process at the DEFINE gate.

Technology is no longer allowed to develop in a vacuum; it is tied to your strategy in a very traceable and measurable format.

The deliverables at the DEFINE gate include the following:

Major deliverables:

- Documented product portfolio requirements (targets and ranges)

- Documented requirements by segments (differences vs. common)

- Documented candidate product portfolio architectures

Supporting DEFINE gate deliverables:

- Segmentation statistics summary

  - Cluster, factor, and discriminate analysis

- VOC-based requirement target and range data

- Common requirements across segments

- Differentiated requirements across and within segments

- Customer survey and validation results

The tasks to be performed within the DEFINE phase that will produce the DEFINE gate deliverables include these:

1. Design and conduct VOC validation surveys to find common vs. differentiated requirements for use in time and circumstance (targets and ranges).

2. Conduct statistical analysis on VOC survey data sets to refine segment identification.

3. Translate NUD customer needs into portfolio requirements (translation into business vocabulary with measurable units).

4. Refine market segments based on VOC validation survey results.

5. Define candidate product portfolio architectures built from mixes of product ideas that meet portfolio requirements across and within segments.

6. Refine market and segment behavioral dynamics maps.

7. Refine technology dynamics maps.

8. Translate product portfolio requirements to R&D.

9. Create EVALUATE phase project plan and risk analysis.

The tools, methods, and best practices that enable the tasks associated with fulfilling the DEFINE deliverables include these:

- Statistical survey design and analysis
- Statistical data analysis
  - Descriptive and inferential statistics
  - Sample size determination, t-tests
  - Multivariate statistical analysis
- Quality function deployment (VOC translation to portfolio requirements)
  - Portfolio Houses of Quality
- Product portfolio-architecting methods
- Project-management tools
  - Monte Carlo simulation (statistical cycle-time design and forecasting)
  - Failure modes and effects analysis (cycle-time risk assessment)

Table 4.2 integrates the Define phase requirements, deliverables, tasks, and enabling tools.

TABLE 4.2   Tools, Tasks, Deliverables-to-Requirements Integration Table

Requirement	Deliverable(s)	Task(s)	Tool(s)
Define the new, unique, and difficult (NUD) product portfolio requirements.	Documented portfolio requirements (targets and ranges) • Documented requirements by segments (differences vs. common) • Segmentation statistics summary from cluster, factor, and/or discriminate analysis • VOC-based requirement target and range data • Common requirements across segments • Differentiated requirements across and within segments • Customer survey and validation results	Design and conduct VOC validation surveys to find common vs. differentiated requirements for use in time and circumstance (targets and ranges). Conduct statistical analysis on VOC survey data sets to refine segment identification. Translate NUD customer needs into portfolio requirements (translation into business vocabulary with measurable units). Refine market segments based on VOC validation survey results.	Statistical survey design and analysis Statistical data analysis Descriptive and inferential statistics Sample size determination, t-tests Multivariate statistical analysis QFD and the House of Quality applied to market/segments needs and their translation into portfolio requirements
Define candidate product portfolio architectures (various mixes of ideas).	Documented candidate product portfolio architectural alternatives	Define candidate product portfolio architectures built from mixes of product ideas that meet portfolio requirements across and within segments.	Portfolio-architecting methods

continues

TABLE 4.2 Tools, Tasks, Deliverables-to-Requirements Integration Table (continued)

Requirement	Deliverable(s)	Task(s)	Tool(s)
		Refine market and segment behavioral dynamics maps.	Process mapping Customer ID matrix Customer interview guides Customer-interviewing methods KJ analysis Questionnaire and survey design and analysis
		Refine technology dynamics maps.	Process mapping Technology roadmapping methods Benchmarking Value chain analysis Key events timelines Patent analysis
		Translate and transfer product portfolio requirements to R&D.	QFD, the House of Quality Requirements documentation methods
Evaluate phase project plan.	Documented Evaluate phase project plan	Conduct Evaluate phase project planning.	Project-management methods Monte Carlo simulation Critical path/chain FMEA

The EVALUATE Phase Tools, Tasks, Deliverables, and Requirements

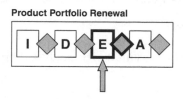

Producing the EVALUATE phase deliverables requires an investment in numerous detailed inbound technical, marketing, and financial-analysis tasks enabled by specific tools, methods, and best practices. Their integration ensures that the right data is being developed and summarized to fulfill the major requirements at the EVALUATE gate review.

The major requirements for the EVALUATE gate include the following:

1. Summary of the candidate product portfolio architectures

2. Best portfolio architecture from among the candidates

   • A hybrid portfolio architecture that contains the best balance from elements of the candidate architectures (collaborative innovation)

3. Financial potential and entitlement from the selected product portfolio architecture for comparison to the overall growth goals for the business

4. Other portfolio evaluation summaries (other than pure financial assessments)

The deliverables at the EVALUATE gate include these:

Major deliverables:

• Documented growth potential of the balanced product portfolio architecture (financial targets and tolerances)

• Documented product portfolio architecture

- Documented risk assessment for the portfolio
  - Balance across multiple dimensions of risk
  - Portfolio FMEA
- Documented Real-Win-Worth analysis across the elements of the portfolio

Supporting EVALUATE gate deliverables:

- Preliminary business cases for the elements in the portfolio
  - Financial models
  - Value propositions
  - Market dynamics and fit with trends
  - Technical risk profiles
  - Competitive benchmarking data and trend analysis
  - Market perceived quality profile and gap analysis
  - Porter's Five Forces analysis and segmented risk profiles
  - SWOT analysis and segmented risk profiles

The tasks to be performed within the EVALUATE phase that will produce the deliverables include these:

1. Create preliminary business cases for elements of the candidate portfolios.

2. Create portfolio evaluation criteria and benchmark portfolio architecture.

3. Conduct portfolio architecture evaluation and selection process.
   - Identify balanced, hybrid portfolio architecture.

4. Conduct portfolio financial assessment (NPV, ECV, ROI, etc.) on selected portfolio architecture.

   • Monte Carlo enhanced portfolio optimization

5. Conduct dynamic rank ordering on selected portfolio architecture.

6. Create ACTIVATE phase project plan and risk analysis.

The tools, methods, and best practices that enable the tasks associated with fulfilling the EVALUATE deliverables include these:

• Financial modeling and forecasting

  • NPV, ECV, ROI analysis

  • Monte Carlo simulation

• Business case development and valuation methods

• Portfolio-balancing methods

• Real-Win-Worth analysis

• Pugh concept evaluation and selection method

• Project-management tools

  • Monte Carlo simulation (statistical cycle-time design and forecasting)

  • Failure modes and effects analysis (cycle-time risk assessment)

Table 4.3 integrates the Evaluate phase requirements, deliverables, tasks, and enabling tools.

TABLE 4.3 Tools, Tasks, Deliverables-to-Requirements Integration Table

Requirement	Deliverable(s)	Task(s)	Tool(s)
Summary of the candidate product portfolio architectures	Documented product portfolio architecture	Create a summary document for each candidate portfolio architecture.	
Best portfolio architecture from among the candidates	Documented hybrid portfolio architecture that contains the best balance from elements of the candidate architectures (collaborative innovation)	Create portfolio evaluation criteria and benchmark portfolio architecture. Conduct portfolio architecture evaluation and selection process. Identify balanced, hybrid portfolio architecture.	
Financial potential and entitlement from the selected product portfolio architecture for comparison to the overall growth goals for the business	Documented growth potential of the balanced product portfolio architecture (financial targets and tolerances)	Conduct portfolio financial assessment (NPV, ECV, ROI, etc.) on selected portfolio architecture. Conduct Monte Carlo enhanced portfolio optimization.	Financial modeling and forecasting NPV, ECV, ROI analysis Monte Carlo simulation
Other portfolio-evaluation summaries (other than pure financial assessments)	Documented risk assessment for the portfolio Balance across multiple dimensions of risk Portfolio FMEA Documented Real-Win-Worth analysis across the elements of the portfolio	Conduct dynamic rank ordering on selected portfolio architecture. Conduct portfolio FMEA. Conduct Real-Win-Worth analysis.	Portfolio-balancing methods Real-Win-Worth analysis FMEA

	Preliminary business cases for the elements in the portfolio Financial models Value propositions Market dynamics and fit with trends Technical risk profiles Competitive benchmarking data and trend analysis Market perceived quality profile and gap analysis Porter's Five Forces analysis and segmented risk profiles SWOT analysis and segmented risk profiles	Create preliminary business cases for elements of the candidate portfolios. Conduct value proposition construction, market dynamics assessment, fit analysis, technical risk assessment, benchmarking, MPQP, Porter's Five Forces analysis, and SWOT analysis.	Business case development and valuation methods Value proposition development Market dynamics mapping Technical risk assessment methods Benchmarking MPQP Porter analysis SWOT analysis
Activate phase project plan	Documented Activate phase project plan	Conduct Activate phase project planning.	Project-management methods Monte Carlo simulation Critical path/chain FMEA

The ACTIVATE Phase Tools, Tasks, Deliverables, and Requirements

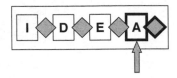

Producing the ACTIVATE phase deliverables requires an investment in numerous detailed inbound marketing and technical tasks enabled by specific tools, methods, and best practices. Their integration ensures that the right data is being developed and summarized to fulfill the major requirements at the ACTIVATE gate review.

The major requirements for the ACTIVATE gate include the following:

1. Rank order of projects to activate

   - Balancing the value of deploying specific commercialization projects at specific timing intervals (when to wait vs. when to take action)

2. Timing for activation of projects based upon a risk-balanced commercialization project-control plan

   - Proper timing and consumption of the company resources and core competencies to safely develop and transfer technology and activate commercialization projects to sustain the growth goals

The deliverables at the ACTIVATE gate include these:

Major deliverables:

- Documented availability, readiness, and deployment of core competencies and resources

- Documented project activation timing and control plan

- Documented FMEA on the product portfolio activation control plan

- Documented enabling technology maturity and readiness

The tasks to be performed within the ACTIVATE phase that will produce the deliverables include these:

1. Rank projects for activation priority and strategic value.

2. Create project-activation timing plan.

3. Conduct risk analysis on the activation plan.

4. Conduct risk analysis on the portfolio for financial performance against growth goals in light of activation timing and planning.

5. Balance resources across the project-activation plan.

6. Generate technology transfer and control plan.

7. Conduct critical parameter management on technologies.

The tools, methods, and best practices that enable the tasks associated with fulfilling the ACTIVATE deliverables include these:

- FMEA methods

- Dynamic rank-ordering methods

- Pareto process

- Resource planning and budgeting methods

- Control-plan design methods

- Technology critical parameter management

- Statistical process control for technologies

- Technology capability studies and assessment

- Project planning and management tools

  - Monte Carlo simulation (statistical cycle-time design and forecasting)

  - Failure modes and effects analysis (cycle-time risk assessment)

Table 4.4 integrates the Activate phase requirements, deliverables, tasks, and enabling tools.

TABLE 4.4 Tools, Tasks, Deliverables–to–Requirements Integration Table

Requirement	Deliverable(s)	Task(s)	Tool(s)
Rank order of projects to activate	Documented project-activation timing and control plan	Rank projects for activation priority and strategic value.	Dynamic rank ordering methods Pareto process
Timing for activation of projects based upon a risk-balanced commercialization project-control plan Proper timing and consumption of the company resources and core competencies to safely develop and transfer technology and activate commercialization projects to sustain the growth goals	Documented availability, readiness, and deployment of core competencies and resources Documented FMEA on the product portfolio activation control plan Documented enabling technology maturity and readiness	Create project-activation timing plan. Conduct risk analysis on the activation plan. Conduct risk analysis on the portfolio for financial performance against growth goals in light of activation timing and planning. Balance resources across the project-activation plan. Generate technology transfer and control plan Conduct critical parameter management on technologies.	Project planning and management tools Monte Carlo simulation (statistical cycle-time design and forecasting) Project failure modes and effects analysis (cycle-time risk assessment) Resource planning and budgeting methods Control plan design methods Technology critical parameter management Statistical process control for technologies Technology capability studies and assessment

Summary of the Major Steps for Product Portfolio Definition and Development Process

1. P&TPR process phases and gates (tools, tasks, deliverables, requirements, project management and performance scorecards for the IDEA process)

2. Market and segment ID (boundaries where profit potential resides for application of our core competencies)

3. Opportunity scoping and mapping (SWOT, core competencies and ideas aligned with market potential), behavior-based process map (internal perspectives, empathy and experiences, use trends and habits, process behaviors)

4. "Over-the-horizon" VOC-gathering for portfolio typing and characterization

   • Interviewing, KJ, NUD screening (market needs within and across segments)

5. Statistical customer survey design and analysis

   • Survey design, t-tests, and customer-ranked NUD requirements (finding statistically significant segments)

6. Product portfolio typing and characterization process (aligning the right type of product portfolio to each significant segment)

7. Product portfolio alternative architecting (portfolio concept generation—generate at least two)

8. Technical platform NUD requirements/technology HOQ/ technology platform architecting (traceable to the product portfolio requirements)

9. Product portfolio valuation and risk analysis

- Real-Win-Worth analysis, FMEA, aggregate NPV, ECV, dynamic rank ordering, other appropriate evaluation tools

10. Product portfolio evaluation and selection

- Pugh process and other concept or architectural-evaluation methods

11. Product commercialization project activation and control plans

- Project management and family plan deployment scheduling, adaptive risk management, resource allocation, and balancing

The IDEA process and its requirements, deliverables, tasks, and enabling tools can add discipline, structure, and measurable results that help ensure that the strategic development and activation of the product and the technology portfolios are aligned and capable of supporting the growth goals of the business. Risk can be better managed with this approach as key strategic decisions are made. Problems are prevented on a proactive basis in technology development and product commercialization with this kind of Six Sigma enablement to portfolio development and renewal.

Tables 4.5–4.7 are the scorecard templates that can be adapted for use during the flow of risk management within and between the phases and gates of the IDEA process.

The Gate Deliverable scorecard (Table 4.5) is used at the gate review at the end of each phase.

The Task scorecard (Table 4.6) is used by the project manager responsible for the detailed tasks conducted within each phase.

The Tool scorecard (Table 4.7) is used by the multifunctional team members who actually apply the tools to fully complete their tasks as they produce the required deliverables.

TABLE 4.5 Gate Deliverable Review Scorecard

1	2	3	4	5	6
Gate Deliverable	Grand Avg. Tool Score	% Task Completion	Results vs. Requirement	Risk Color Code(R-Y-G)	Gate Requirement

TABLE 4.6 Task Scorecard

1	2	3	4	5	6
Task	Avg. Tool Score	% Task Completion	Task Results vs. Required Deliverable	Risk Color Code (R-Y-G)	Gate Deliverable

TABLE 4.7 Tool Scorecard

1	2	3	4	5	6
Tool	Quality of Tool Use	Data Integrity	Results vs. Required Task Deliverable	Specific Tool Score	Task & Deliverable

Refer to Chapter 2, "Scorecards for Risk Management in Technical Processes," for the details on how these scorecards are to be used.

5

STRATEGIC RESEARCH AND TECHNOLOGY DEVELOPMENT PROCESS

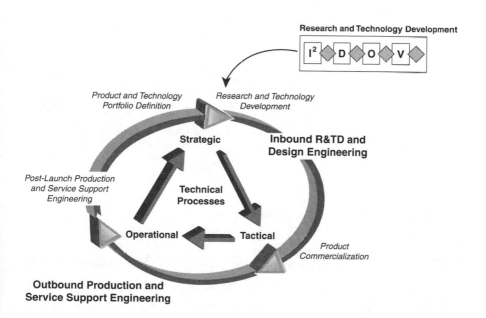

Six Sigma–Enhanced Research and Technology Development

The development of new technology is the fuel that powers the timely delivery of new products and new production processes. It enables the fulfillment of the product portfolio requirements and strategic growth goals of the business. Within this portion of the total product development process, advances in performance and quality are actually born. Any company that does not view its R&TD team as the strategic starting point (see Figure 5.1) for the physical development of quality and for laying the foundation for efficient cycle-time in product commercialization is very much off base. Six Sigma in R&TD prevents downstream problems.

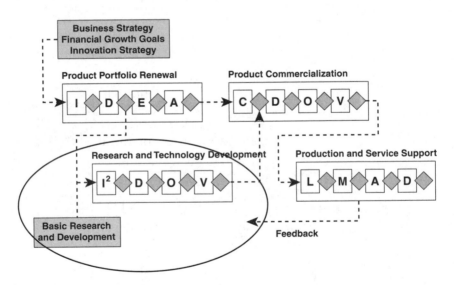

FIGURE 5.1   Integrated process diagram.

If the R&TD work is not properly completed before transfer to a commercialization project, we are left with what we call technology WIP (work-in-process). Technology WIP is a leading cause of late launches and cycle-time corruption during commercialization. Is all technology WIP bad? Not always—it depends on the nature of the risk the incomplete work carries into a given commercialization project. Artificial deadlines create artificial truncation of tasks. This is what really kills growth potential, because everyone works in chaotic workflows that cannot produce complete, disciplined results that underwrite the efficiency and productivity we expect from our product-development pipeline.

We consider research and technology development a strategic process that is heavily influenced by the product portfolio renewal process. Product commercialization is a tactical process that is fed by these two strategic processes. Product portfolio renewal activates a product commercialization project, while R&TD transfers robust and tunable technologies that enable product commercialization projects. This chapter takes the position that the teams within research and technology development organizations have the best and earliest opportunity to develop the foundations for Six Sigma performance within the detailed processes used to generate new and leveraged technologies. So if you were to ask the question, "Can some Six Sigma methods be deployed in the day-to-day activities within an R&TD process?" the answer is, "Absolutely, yes."

This chapter constructs an architecture for the deployment of such methods in the development and transfer of new and leveraged technologies. It also illustrates how the strategic product portfolio renewal team delivers its requirements to the R&TD community so it can begin strategically focused technology development projects. Unique here is that R&TD rarely gets specific functional or performance targets to hit and is much more likely to get ranges of performance that can be adjusted as the clarity of application increases when specific opportunities and their embryonic business cases emerge at the Activation phase of the IDEA process. We seek technology that can perform across ranges because we simply do not know what the specific targets are yet.

A Six Sigma project in R&TD is the development of a new technology as a complete, integrated entity to be transferred to a product commercialization project. Industry is now beginning to recognize that technology-development projects should be largely independent from product commercialization projects. This is never an easy thing to do, but if you want to minimize technical difficulties and schedule delays as you approach product design verification and launch during commercialization, this is precisely what must be done. It is very risky behavior to develop new technology within a product commercialization project. Technology WIP in a commercialization project must be carefully managed. If you want to conduct "lean" commercialization, you must minimize—and, in some unique cases, eradicate—technology WIP.

Big differences exist between what tasks need to be done (work) and what facts need to be known (data) during technology development and product commercialization. Lean and Six Sigma have an impact on value-added work and the trustworthiness of data. Technologies that are under development are immature and overly sensitive to sources of manufacturing, distribution, and customer use variability (known as noise factors). When they are prematurely forced into the high-pressure timeline of a commercialization project, corners get cut (work is truncated) and sensitivities to variation go undetected. Immature, underdeveloped technologies can greatly slow the progress of the product due to their inability to integrate smoothly and function correctly in the presence of noise factors (a noise factor is any source of variation from external, unit-to-unit, or deterioration sources). In a word, they are not "certified" for safe transfer to commercial or design application. Deploying new technologies before their certification is analogous to launching a new product design that has not been made "capable" from a manufacturing, assembly, and serviceability perspective. To the statistician, it is similar to running a process before one has proven it to be under a state of statistical control. You must quantitatively prove that a new technology is both robust and tunable before it can be used in a new product. By "robust," we mean capable of short-term performance as measured by Cp indexes, even in the presence of special-case noise factors. By "tunable," we mean capable of long-term performance as measured by Cpk indexes. Tunable technology has the designed-in capability to adjust the mean functional performance to hit any forecasted target across a range of performance for a family of products. This then makes k = 0 and causes Cpk to equal Cp.

This chapter lays a foundation for a process to properly develop and transfer new technologies into a product commercialization project. Some have said that Lean and Six Sigma are not applicable to R&TD. We hope this paragraph makes a compelling case for the appropriateness of Lean and Six Sigma as entities that can indeed add value during R&TD. If you get nothing else from this text, remember that we are trying to design an R&TD process that is built on customer-oriented, value-added workflow (lean stuff) and the transfer of robust and tunable technologies that are desensitized to strong sources of variability (Six Sigma stuff).

Technology development and transfer must be properly controlled and managed by facts generated using a system-integration process we call critical parameter management.

The proper place to initiate critical parameter management in a business is during advanced product portfolio planning and R&TD. If critical parameter management is practiced within these earliest stages of product development, a certified portfolio of critical functional parameters and responses can be rapidly transferred as a "platform" or family of modular designs into the design of families of products during numerous product commercialization programs. This will efficiently help you reach a steady state of growth.

Critical parameter management is a systems-engineering and integration process that is used within an overarching technology development and product commercialization roadmap. We first look at a roadmap (I^2DOV, pronounced "eye-dov") that defines a generic technology-development process and then moves on to define (in Chapter 6, "Tactical Product Commercialization Process") a similar generic roadmap (CDOV, pronounced "see-dov") that defines a rational set of phases and gates for a product-commercialization process. These roadmaps can be stated in terms of manageable phases and gates. These generic phases and gates are useful elements that help communicate how to conduct product development with order, discipline, and structure. You should use your own phase-gate vocabulary. Our IIDOV (sometimes referred to as I^{2}DOV) vocabulary is just a "word bridge" to help you integrate Six Sigma discipline into your processes.

The I^{2}DOV Roadmap: Applying a Phase-Gate Approach to Research and Technology Development

Research and Technology Development

The structured and disciplined use of phases and gates to manage product commercialization projects has a proven track record in a large number of companies around the globe. A small but growing number of companies are using phases and gates within their R&TD organizations to manage the development of new and leveraged technologies *prior to commercial design application*. In the next two chapters, we show the link between the use of an overarching R&TD and product-commercialization process; deployment of a full portfolio of modern tools, methods, and best practices; and a well-structured approach to project management from R&TD all the way to product launch (see Figure 5.2).

With this triangle of structural stability, one can drive for breakthroughs that deliver designed results because portfolio, technology, and commercialization teams all know what to do and when to do it.

These three stabilizing forces make it work. If you are wondering why major redesign initiatives of your phase-gate process seem to falter and fail to create a sustainable way of conducting efficient product development, you are probably missing disciplined integration of the other two legs of the triangle. Functional excellence in project management and tools, methods, and best practices are a must for any phase-gate process to actually work and have any true credibility. If your people think your phase-gate process is a joke, you need to enlist some of them who are respected by their peers to structure it so that everyone respects it and makes an honest effort to live by its discipline. This triangle will go a long way toward getting your process to a level of respect and use by your teams. When you have this in place, you have to let your teams use it without interference or optionalism. Either we do our work thoroughly and with craftsmanship or we should cancel the project and focus on the few we will commit to doing right. Portfolio process excellence means we are committing to doing the right few projects. R&TD and product-commercialization process discipline means we will do those few projects right. When we select our strategic few projects, we should hurry along and get them done right—but we should never overcommit our teams to take on too much work and then force them to rush. This is a recipe for inconsistent growth. So if you want sinusoidal growth patterns, do too many projects with too few people—and, on top of that, make them rush their tasks.

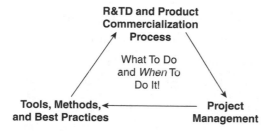

FIGURE 5.2 Triangle of process excellence.

No one R&TD process architecture will meet the needs and priorities of all companies. Every company must define a process that fits its own culture, business process, and technology use paradigm. Lets move on and review the details of the phases and gates of the IIDOV process. As we do this, we begin to develop the critical parameter management context that is helpful in the development of Six Sigma performance (or any desired "sigma," for that matter).

The IIDOV phases and gates structure (see Figure 5.3) for R&TD outlined in this chapter follows four discrete project-management steps:

Phase 1.   II = Invention and innovation

Phase 2.   D = Develop

Phase 3.   O = Optimize

Phase 4.   V = Verify

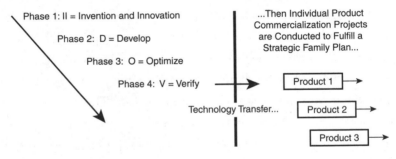

FIGURE 5.3   IIDOV process phases.

We define each of these phases and describe how they flow as part of the overarching product-development process that leads to successful product launches.

I²DOV Phase 1: Invent/Innovate Technology Concepts

Research and Technology Development

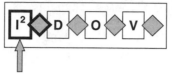

Technology requirements and concept definition.

The Invent/Innovate gate requirements include the following:

- Defined technology roadmaps and trends

- Technology platform and modularity requirements

- Technology concepts

- Technology risk profiles

- Develop phase project plans

The Invent/Innovate gate deliverables include these:

- Documented technology roadmaps

- Summary of trends in technology dynamics in our strategic arenas

- Documented technology concepts (platforms and modules)

- Documented risk profiles for our technology concepts

- Documented project plans for the projects moving into the Develop phase

The Invent/Innovate tasks include these:

1. Construct technology roadmaps and document technology trends.

2. Gather and translate "over-the-horizon" voice of the customer.

3. Define product line strategies (PLS) and family plans.

4. Create technology Houses of Quality.

5. Conduct technology benchmarking.

6. Generate technology system (platform or modularity) requirements document.

7. Define new, unique, and difficult functions that fulfill the technology requirements.

8. Enable, support, and conduct invention (generate new knowledge).

9. Enable, support, and conduct innovation (enhanced/leveraged reuse of existing knowledge and technologies).

10. Refine, document, and link discoveries, learning, and basic models to product line strategy.

11. Transfer refined, linked, and documented knowledge into technology concept design workflow.

12. Generate technology concepts that fulfill the new, unique, and difficult functions.

13. Generate risk assessment and summary profile.

14. Generate Develop phase project plans and cycle-time models.

This flow of 14 tasks can be conducted with the help of a set of tools, methods, and best practices under the timing constraint of the principles of project management. With Six Sigma–enabled R&TD, a technology manager typically can construct a PERT chart, conduct Monte Carlo simulations, and conduct a cycle-time FMEA to plan the timing of these tasks as they relate to use of a balanced portfolio of tools, methods, and best practices.

I^2DOV Phase 1: Invention and Innovation Tasks

This section reviews the specific issues related to each action item. Each of these tasks can be conducted by using a tool, method, or best practice. The flow of tasks can be laid out in the form of a PERT chart for project-management purposes.

We generally view the Invention and Innovation phase as the beginning of technology development. It is also the beginning of the critical parameter-management process. Before any technology development work can begin, the requirements of the business must be well-defined. The businesses goals, strategy, and mission need to be clearly established. The corporate leadership team must document short-term and long-term business and financial targets that constitute its goals. The team must have a documented plan that will achieve the goals. This constitutes the team's strategy. The leaders of the business must work with the board of directors to document a mission statement that defines core values that provide boundaries for the businesses strategic framework and goal-setting process. Everyone must understand why the business exists.

Here the market and technology teams discover where to gather the longer-term, "over the horizon" voice of the customer. Defined markets and market segments are essential to guiding the advanced product-planning team as they go out to discover what customers need so they can apply their core competencies properly. When it is clear what market segments exist, the advanced product-planning team can conduct more specific market focused activities for gathering the voice of the customer. Technologies can be developed to apply across or within market segments. It is important to define where the technology needs are synergistic across market segments because this will clearly signal to the R&TD leaders that a major platform opportunity exists. If a technology platform can be developed to serve "across-market" commercialization requirements, great economies of scale can be attained. This scenario leads to the single investment of capital that can deliver profit from numerous family lines of products. To a lesser extent, profit amplification can occur within a market segment by developing a technology platform that can serve "within-market" product family plan requirements. Market segmentation is a great help in product family planning. Without it, companies don't know when or how to divide R&TD resources to help attain market share and when to integrate resources to attain economies of scale in the utilization of their R&TD resources.

Construct Technology Roadmaps and Document Technology Trends

Technology roadmaps are concurrently developed with the market-segmentation documents. They help the business and technical leaders structure what technology development projects should be undertaken during the first phase of technology development.

Technology roadmaps document past and current technology patents, cost structures, performance, and development cycle-times. They also develop future technology forecasts within and outside the company. A good technology roadmap contains the following three items:

- **Historical technologies** that have completed their lifecycles. They should contain documented development cycle-times, costs vs. generated revenue streams, and the details of their functional performance during their lifecycles.

- **Current technologies** that have not yet completed their life-cycles. Their performance, cost, and development cycle-times should all be documented.

- **Future technologies** should be scoped for estimated cost, development cycle-time, and functional performance capabilities. These projections should be based on both technology trend forecasting and the long-range voice-of-the-customer assessment. Latent customer needs are estimated within this form of documentation.

Gather and Translate "Over-the-Horizon" Voice of the Customer

The use of voice-of-the-customer data is essential for circumspect advanced product planning. Pure invention and innovation that is not in some way linked to customer needs can produce technology that is protected by a deep portfolio of uncommercializable patents. This is referred to as *design for patentability*. If your company is not careful, it can produce what is called a 6-foot-deep patent portfolio. This occurs when unbridled R&TD organizations develop technology independent from the inbound marketing group and the advanced manufacturing organization. The company becomes rich in patents that are incapable of being successfully transferred into revenue-bearing commercialization projects. A good question to ask about your company's patent portfolio is, "What is the revenue-bearing product hit rate based on the patent portfolio?"

It is essential that the technologists participate in gathering, ranking, and translating customer needs as part of the advanced product/technology planning process. This helps them be circumspect as they seek balance between their internal inventive activities and the development of technologies that fulfill explicit, external market needs. On the other hand, it is entirely possible to follow the voice of the customer into a business failure (see *The Innovator's Dilemma*, by Clayton M. Christenson). Invention of new technologies that surpass known customer needs (*often called "latent needs"*) often plays a key role in new product development.

Part of the technology-development process includes the invention and discovery of unique phenomena that could be refined into a safe and mature technology, independent of any known customer need. The focus in this case is on "the observation, identification, description, experimental investigation, and theoretical explanation of natural phenomena." This is the definition of science (from the *American Heritage Dictionary*) and largely embodies the process of invention. The voice-of-the-customer data might or might not play a strong role in the "discovery of science" process. Scientists and engineers who participate in the invention process might use VOC data to guide their thinking, but many do not like any constraints on their thought processes. Really only two choices exist during technology development: First is to recognize market needs and then drive the inventive processes so that the discovery of physical law is directly targeted to be commercially useful according to an advanced product-planning strategy. Second is to just invent in an unrestrained fashion, and if what you come up with is somehow commercially useful, bravo—you have good luck!

Define Product Line Strategies (PLS) and Family Plans

The long-range customer needs are initially defined in broad terms and are then refined into product line technology requirements based on market segmentation. The R&TD organization should help gather and process this customer information. After the long range VOC data is used to define market segment needs, additional VOC data can be acquired to develop specific product line or family requirements. R&TD teams use documented product-line requirements as the basis for developing platform technologies (system and subsystem technologies that can be used as a basis for a family of products). The platforms are intended to fulfill the voice-of-the-customer requirements over a family of commercial products. Requirements for individual technologies that might or might not support a platform are also identified and documented. Unfortunately, family planning that is tactically fulfilled by a technology development process is largely nonexistent in many modern companies.

Create Technology Houses of Quality

The VOC data for a specific market segment can be used to document a product line strategy and a series of embryonic family plans. The specific needs projected by the long-range VOC data and the technology roadmaps will provide the balanced input needed to construct a series of technology Houses of Quality. The balance comes from the integration of the voice of the customer and the voice of technology.

A technology House of Quality is created using a matrix approach while translating technology *needs* into technology *requirements*. After a group of technology requirements are defined for a family of products, the R&TD team can begin the highly technical task of defining and developing physical functions that fulfill the requirements. These physical functions lie at the heart of conceptual development of technology architectures that can ultimately become a platform. A technology House of Quality is the foundation upon which a platform is conceived and architected.

Conduct Technology Benchmarking

A major portion of a technology House of Quality depends on three forms of comparative information:

- Projected customer opinions and perceptions of existing and new trends in technology that will drive fulfillment of their needs in product performance.

- Projected technology progressions based on current and foreseeable technological innovations. Frequently, these are forecasted in the latest patents and invention disclosures.

- Current technology benchmarking that assesses the state-of-the-art in technology being applied in new products that are currently on the commercialization scene.

The technology House of Quality uses the projections and benchmark data to help assign priorities and rank criticality during the translation of technology needs (family plan needs) into technology requirements ("hard" technical metrics associated with need fulfillment).

Generate Technology System Requirements Document

When the technology House of Quality is built and filled out, it forms the basis for the development of a formal document that contains the technology requirements, which the appropriate teams of technologists within the R&TD organization will fulfill. Not all requirements have to come from the technology House of Quality. The HOQ typically focuses attention on those things that are *new, unique, or difficult*. Some technology requirements might fall into less demanding categories and obviously still need fulfillment even though they were not a point of focus from the perspective of the HOQ. For example, a technology might need to fulfill some readily known and easily fulfilled regulatory or safety requirements. These requirements need to be integrated along with the new, unique, and difficult requirements that flow from the HOQ.

A new system of technology made up of integrated subsystem technologies is called a *platform*. The platform is considered a system. Platform development is well-served by approaching it as one would any engineered system. Here we see very strong similarities between a system or product requirements document and the platform requirements document. In *Design for Six Sigma for Technology and Product Development*, an entire chapter is dedicated to the detailed process of system engineering in the context of DFSS.

If you are focusing the technology development activity on just one or more subsystems, the requirements document is still developed—but just for the subsystem technology. In both cases, a balanced portfolio of tools and best practices is used to fulfill these requirements.

Define Functions That Fulfill the Technology Requirements

The real technical work of the R&TD scientists and engineers begins at this step. Here actual physical and engineering principles are identified to fulfill the requirements. Specific functions are described to make clear the underwriting physics that define how the requirements will be fulfilled. This is the time for the technology-development teams to define what will have to be "made to happen" to enable the function behind the technology. The actual form and fit concepts come later during the final stages of concept engineering. Here we are talking about functions, *independent of how they are attained.* A force might be required—an acceleration, a displacement, a reaction ... whatever dynamic transformation, flow, or state changes of mass, energy, and information (analog or digital logic and control signals) are necessary to satisfy a requirement. When the team understands the nature of the fulfilling functions, it can enter into conceptual development with real clarity of purpose. It can innovate freely to develop multiple architectures ("forms and fits") to fulfill the functions that, in turn, fulfill the requirements.

Enable Support and Conduct Invention (Generate New Knowledge)

If a requirement demands a function that is not found in the current physics or engineering knowledge base, a new invention might be required to meet the need. Far more commonly, innovation is used to integrate and extend applications of known physical principles and engineering science. It is extremely important to note that the risk associated with this set of activities forces us to recommend that you stay clear of conducting technology development within product commercialization processes. When you get to the point of a required invention, it is very likely that developing the right function might be extremely difficult or impossible. Invention can burn up a lot of time that you simply do not have in a product commercialization project that is tied to a launch date. When this happens to a "design" that is really just an uncertified technology that has been prematurely inserted into a commercialization project, the project schedule is usually compromised and profound implications arise for meeting the market window of the product. This form of "untamed technology" holds the impending revenues hostage. Our recommendation is to avoid mixing uncertified technologies into your product commercialization projects.

Enable, Support, and Conduct Innovation (Enhanced/Leveraged Reuse of Existing Knowledge)

Innovation is by far the most popular form of developmental activity that characterizes R&TD processes as we enter the twenty-first century. Some call it applied research and development. The operative word is *applied*. That means a pre-existing technology knowledge base is available for reuse. It is documented knowledge that can be applied either by using it in a novel way or by extending or advancing the state-of-the art. Most Baccalaureate- and Master's-level engineers are trained to extend the state-of-the-art, while most PhD-level engineers and scientists are prepared to create and discover new knowledge. To some, this might not be a very clear point of distinction, but for those of us who have been in a situation in which time is of the essence, invention of new knowledge is simply not a timely option. We must use what is known to get the technology developed to meet the needs of our colleagues in product design.

Refine, Link, and Document Discoveries, Learning, and Basic Models to PLS

After new knowledge and reused but newly extended knowledge have been developed, it is time to complete the construction of a very important bridge. That bridge links what the team has documented as a knowledge base to a set of well-defined application requirements related to the family of products that will flow from the product line strategy. This is how we integrate the practical results being born out of the technology strategy with the product line strategy, ultimately fulfilling the business strategy.

The long-range voice of the customer and our benchmarked voice of technology lead us to construct a technology House of Quality, which leads us to a technology requirements document, which leads us to define technology functions that enable the fulfillment of these diverse needs. The knowledge of "functionality fulfillment" must be held up to the product line strategy in a technology review to answer the questions related to "Can we get there from here?" This test of feasibility comes before investing a lot of time and resources in concept generation. The business leaders should see a rational flow of fulfillment at this point. If it is not clear, further work needs to be done to align all the fulfillment issues with the product line strategy.

Transfer Refined, Linked, and Documented Knowledge into Technology Concept Design

All the knowledge now must be moved forward into a format that enables the next stage of technology development, known as technology concept development. The functions and requirements must be documented so that a rational set of concept-generation, evaluation, and selection criteria can be rapidly developed. If ambiguity exists regarding what the concepts must do and fulfill, the concepts will inevitably suffer from competitive vulnerability and will be functionally inferior. Concept selection is yet another test of feasibility, but this time it has to do with both functional and architectural potential to fulfill the technology requirements.

Generate Technology Concepts That Fulfill the Functions

The technology team now focuses the full weight of its individual and corporate development skill to generate candidate architectures to fulfill the functions and requirements that have been tied back to the product line strategy.

For integrated subsystems of technologies, the goal of the team is to first define a number of candidate system architectures known as platforms. Platforms are used to produce a number of products from one investment of capital within the R&TD organization. Platforms are necessary to make efficient use of business capital to fulfill product line strategies. You will find it very hard to fulfill long-range business strategies through R&TD projects that deliver one subsystem at a time for refitting onto aging products. A company might use a mixed-technology development strategy of delivering some new subsystems to extend existing product families while new platforms are on the way. Eventually, however, a new platform must emerge to sustain the business as markets and technologies evolve. In some business models, technology turnover is so rapid that the notion of a reusable platform does not apply.

Numerous candidate subsystem or platform concepts need to be generated. If a lone concept dominates from one strong individual or "clique" within the R&TD organization, a small likelihood exists that the company will identify the best possible concept. The entire team should break up and as individuals generate several concepts. After many concepts are generated, the team can convene to evaluate and integrate the best ideas into a superior "superconcept" that is low in competitive vulnerability and high in functional feasibility. The technology-development team can come to a consensus on this state of superiority using concept-selection criteria developed from the previously developed functions and requirements.

Generate Risk Assessments and Summary Profile

Every phase of the R&TD process should culminate in a summary assessment of accrued risk. This is enabled by data from the tool-task clusters conducted within each phase. When summarized, this data is referred to as a deliverable. Our knowledge and clarity of risk builds phase by phase. In fact, we structure our work into phases so we can group accrued risk in manageable sets so we can make decisions at known points of concern and investment control.

Risk is best identified by evaluating deliverables against their requirements. If your requirements are unclear and anemic, your ability to assess risk will be, too. If deliverables are not underwritten by disciplined completion of tasks enhanced by proper tools, methods, and best practices, you will have a very hard time making good decisions at each gate review.

Generate Development Phase Project Plan

Chapter 2, "Scorecards for Risk Management in Technical Processes," fully describes this, but it is worth mentioning at this point. Every gate review should focus 80 percent of its time on the thorough review of the requirements, deliverables, tasks, and tools that will be in play during the next phase of work. Each R&TD leader should be at least one phase ahead of the teams that are doing the work. This prevention strategy is considered a key competence for all managers working in a phase-gate context. A good project plan has a clear summary of the work break-down structure, a projection of the critical path, a Monte Carlo simulation of alternative critical-path scenarios, and a cycle-time FMEA for the major tasks on the critical path. Anything less proves your management team does not have a realistic view of what likely will happen during the next phase of work. With this kind of plan in hand one phase ahead, the management team can become very proactive in preventing problems that they can influence.

GATE 1 Readiness

At this early point in technology development, a business should have a proprietary, documented body of science; documented customer needs for future product streams; and the embryonic technical building blocks that can be developed into candidate technologies. The term *fuzzy front end* is appropriate to characterize the nature of this phase of R&TD. We don't have mature technologies yet—just the science, customer needs, and candidate concepts from which we hope to derive them.

With a viable cache of candidate technical knowledge and a well-planned commercial context in which to apply that knowledge, we approach a rational checkpoint to assess readiness to move into formal technology development.

Whether technology is built on completely new knowledge or integrated from existing knowledge (something we call leveraged technology), critical parameter relationships must be identified, stabilized, modeled, documented, desensitized to sources of variation, and integrated across functional and architectural boundaries. Many of the delays that are incurred during product commercialization result from technologies that have incomplete understanding and a lack of documentation of critical parameter relationships. As far as this text is concerned, no distinction is made between newly invented technologies and innovations that reuse existing technologies. Both situations require structured, disciplined R&TD teamwork to safely deliver "certified" technologies to the commercialization teams. When a certified technology is in the hands of a properly staffed and trained commercialization team, the team can rapidly transform it into capable designs. These issues must be foremost in the minds of the decision makers as the technology-development process flows through its phases and gates.

At Gate 1, we stop and assess the candidate technologies and all the data used to arrive at them. The tools, methods, and best practices of Phase 1 have been used to fully complete the key tasks. Their deliverables and summary results can be summarized in a Gate 1 scorecard for risk assessment. A checklist of Phase 1 tools, tasks, and their deliverables can be reviewed for completeness of results and corrective action in areas of risk—all in light of the Phase 1 requirements.

A General List of Phase 1 Tools, Methods, and Best Practices

- Technology roadmapping
- VOC gathering methods

- KJ analysis

- QFD/House of Quality—Level 1 (system level)

- Patent analysis and exposure assessment

- Market segmentation analysis

- Economic and market trend forecasting

- Technology business case development methods

- Competitive benchmarking

- Scientific exploration methods

- Knowledge-based engineering methods

- Math modeling (business cases, scientific and engineering analysis)

- Concept-generation techniques (TRIZ, brainstorming, brain writing, and so on

- Modular and platform design

- System-architecting methods

- Technology FMEA

- Project-management methods

- Monte Carlo simulations

- Project FMEA

Table 5.1 integrates the Invent/Innovate phase requirements, deliverables, tasks, and enabling tools.

TABLE 5.1   Tools, Tasks, Deliverables–to–Requirements Integration Table

Requirement	Deliverable(s)	Task(s)	Tool(s)
Defined technology roadmaps and trends	Documented technology roadmaps Documented summary of trends in technology dynamics in our strategic arenas	Construct technology roadmaps. Document technology trends.	Technology roadmapping Patent analysis and trend assessment
Technology platform and modularity requirements	Documented technology requirements	Gather and translate "over-the-horizon" voice of the customer. Define product line strategies (PLS) and family plans. Create technology Houses of Quality. Conduct technology benchmarking. Generate technology system (platform or modularity) requirements document.	VOC-gathering methods KJ analysis QFD/House of Quality—Level 1 (system level) Market-segmentation analysis Economic and market trend forecasting Technology business case development methods Competitive benchmarking
Technology concepts	Documented technology concepts (platforms and modules)	Define new, unique, and difficult functions that fulfill the technology requirements. Enable, support, and conduct invention (generate new knowledge). Enable, support, and conduct innovation (enhanced/leveraged reuse of existing knowledge and technologies). Refine, document, and link discoveries, learning, and basic models to product line strategy. Transfer refined, linked, and documented knowledge into technology concept design workflow. Generate technology concepts that fulfill the new, unique, and difficult functions.	Scientific exploration methods Knowledge-based engineering methods Math modeling (business cases, scientific and engineering analysis) Concept-generation techniques (TRIZ, brainstorming, brain writing, and so on Modular and platform design System-architecting methods

| Technology risk profiles | Documented risk profiles for our technology concepts | Generate risk assessment and summary profile. | Patent analysis and exposure assessment Technology FMEA |
| Develop phase project plans | Documented project plans for the projects moving into the Develop phase | Generate Develop phase project plans and cycle-time models. | Project-management methods Monte Carlo simulations Project FMEA |

I^2DOV *Phase 2: DEVELOP Technology*

Research and Technology Development

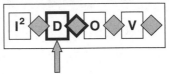

Technology concept definition, stabilization, and functional modeling.

The Develop gate requirements include the following:

- Superior technology concepts

- Documented math models, prototypes, and measurement systems

- Baseline functional stability, tunability, and capability under nominal conditions

- Technology risk profiles

- Optimize phase project plan

The Develop gate deliverables include these:

- Documented superior technology and measurement system concepts that are worth taking into the technology-optimization phase

- Documented underlying math models (deterministic and statistical) that characterize the nominal and tunable performance (baseline) of the new technology concepts and their measurement systems.

- Prototype hardware, firmware, and software that provides a physical model of the new technology subsystem concepts and their measurement systems.

- Documented SPC charts (stability and tunability); capability indexes for all critical functional responses

- Documented risk assessment

- Documented optimization project-management plan

The Develop phase tasks include these:

1. Generate and refine technology concept evaluation criteria.

2. Evaluate and select superior technology concepts.

3. Analyze, characterize, model, and stabilize nominal and tunable performance of superior technology concepts (models, prototypes, and measurement systems).

4. Conduct SPC and capability studies under nominal conditions to certify and prepare superior technology models and prototypes to take into optimization.

5. Generate risk assessment and summary profile.

6. Generate Optimize phase project plan and cycle-time model.

The nature of this second phase of technology development is one of characterization of nominal and tunable performance relationships between input or independent (x) variables and output or dependent (Y) variables. This phase requires all three major forms of modeling to be conducted:

- Analytical or mathematical models

- Graphical and spatial models (2D and 3D sketches, layouts, and CAD files)

- Physical models (Hardware, firmware, and software prototypes)

In standard Six Sigma terminology, this is the point in the process where the technology-development team quantifies the (Y) variables as a function of the (x) variables. In the vocabulary of critical parameter management, the (Y) variables are called critical functional responses and the (x) variables are candidate critical functional para-meters. Sometimes the (x) variables are not totally independent. At times (x) variables have co-dependencies and/or correlated effects on a (Y) variable existing between them. The former are referred to in statistical jargon as *interactions*. Additionally, the (x) variables might not have a linear relationship with the (Y) variable(s). It is important to be able to quantitatively state whether the relationship is linear or nonlinear and to describe the extent of the nonlinearity, if proven to exist. Finally, we need to know which (x) variables can adjust or tune the mean value of (Y) without greatly affecting the standard deviation of the (Y) variable. The statistical variable used to quantify this rela-tionship is the coefficient of variation (COV): $(\sigma/y_{bar}) \times 100$. The COV is also the basis for a very important measure of robustness (the sig-nal-to-noise ratio) during the optimization phase.

The (x) variables are often divided into two distinct categories: control factors and noise factors. Control factors are (x) variables that the technology-development team can readily control and specify. Noise factors are (x) variables that the engineering team cannot easily control or specify. Sometimes the (x) variables are impossible to con-trol. This phase of technology development provides the "first cut" at distinguishing control factors from noise factors. The (x) factors that are found to be marginally controllable or extremely difficult or expensive to control can be isolated for special attention. If the situa-tion is bad enough, the technology concept might have to be refined or abandoned. If the concept is found to be nominally stable and tun-able, the team can treat these poorly-behaved control factors as noise factors during the optimization phase of technology development.

The multifunctional team that initiates technology concept devel-opment is typically made up of some or all of the following personnel:

- Scientists from various technical domains

- Development engineers from various technical domains

- Advanced manufacturing engineers

- Service specialists

- Product-planning and inbound marketing specialists

- Supply chain development specialists

- Health, safety, environmental, and regulatory specialists

- Technology-development partners from industry, academia, government, and so on

- Selected customers

Generate and Refine Technology Concept-Evaluation Criteria

Concepts must be evaluated against a set of criteria that is directly related and traceable to the technology requirements. In fact, the concept-evaluation criteria can literally be the technology requirements, with little or no refinement from the statements found in the technology requirements document. It is perfectly acceptable to develop the evaluation criteria before generating the concepts. Whether to use the technology requirements document or the typically broader and more detailed evaluation criteria to develop technology concepts is up to the team.

Concept evaluation criteria consists of all key requirements that the technology concept must satisfy. These criteria must comprehensively represent VOC, VOT, and product line strategy requirements. One of the reasons KJ analysis (affinity diagramming) is so important is that it helps state VOC needs and VOT issues in a hierarchy of relationships and in a rank order of importance. This information helps the technology-development team form a comprehensive list of *critical* evaluation criteria. All concepts can be evaluated against a "best-in-class" benchmark Datum Concept. Not uncommonly, one or more of the new candidate concepts bring new criteria to the forefront. This new criteria springs from innovation and newness embodied in the candidate concepts. Care should be taken not to miss this new criteria that helps discriminate and refine conceptual feasibility during the Pugh concept-evaluation process.

Evaluate and Select Superior Technology Concepts

At this point, the Pugh concept-selection process is used to evaluate the set of candidate concepts against the Datum Concept. The Pugh process uses an iterative "converging–diverging" process to evaluate concepts. The ranking of each candidate concept vs. the datum is non-numeric and uses simple (+) "better than the datum"," (–) "worse than the datum," and (S) "same as the datum" ranking designations against the evaluation criteria. The iterative "converging–diverging" process is used to weed out weak concept elements, enabling the integration of hybrid superconcepts from the remnants of candidates that have strong elements against the datum. Iterative "pulses" of concept screening and synthesis build up a small, powerful set of newly integrated hybrid concepts that typically have the capacity to equal or surpass the datum relative to the superior fulfillment of the evaluation criteria.

Analyze, Characterize, Model, and Stabilize Nominal and Tunable Performance of Superior Technology

When a superior concept or two have been defined from iterations through the Pugh concept-selection process, the team is ready to conduct an extensive technical characterization. It is not a bad idea to have a back-up concept as a safety valve in case the superior concept possesses a fatal flaw that was missed during the Pugh process. The superior concepts are proven to be low in competitive vulnerability and high in technological superiority only relative to the datum. They are worth the investment required to certify the new technology's nominal and tunable performance before optimization. Once in a while, a superior concept passes the concept-evaluation process only to be found unusable under the deeper analytical scrutiny of this next step in the development of the technology. As you can see, Phase 2 is very important for weeding out technologies that are often "time bombs" when allowed to creep into a product commercialization process without proper development and scrutiny. Conducting failure modes and effects analysis on the superior concepts at this point is extremely valuable in exposing risks.

The analysis and characterization must include the development of all key "Y as a function of x" relationships. As previously mentioned, each relationship must be quantified in terms of nominal and tunable performance. Linear and nonlinear relationships must be defined. Interactions between control factors must be identified. If these relationships are a problem, the interactive relationships must be re-engineered so that the codependency between two control factors will not cause severe variation-control problems later in the product commercialization process. This re-engineering to survive interactivity between control factors (called *design for additivity*) can first be approached in Phase 2 of technology development and finalized in the Phase 3 optimization activities. In Phase 3, we add the extra characterization of interactions between control factors and noise factors (the main focus of robust design).

The most important knowledge that must come from this phase of technology development is the quantification of the ideal/transfer functions that underwrite the basic performance of the technology. In the next phase, we evaluate the controllable (x) variables that govern the performance and adjustability of the ideal/transfer function for their robust set points in the presence of the uncontrollable (x) variables known as noise factors. If we have incomplete or immature forms of the "Y is a function of x" relationships and fail in our attempt to attain robustness, we can pass along yet another time bomb for our peers in product commercialization. By *time bomb*, we mean a technology that gets integrated into a product system and "blows up" by producing unacceptable or uncontrollable performance with little or no time left to correct the problem. Technology that is poorly developed and characterized is a major reason why product delivery schedules slip.

Conduct SPC and Capability Studies under Nominal Conditions to Certify and Prepare Superior Technology Models and Prototypes to Take into Optimization

It is challenging to know when a team has enough knowledge developed to state that a new technology is truly ready to go forward into the optimization phase of technology development. Let's look at a list of evidence that could convict us in a court of law of having conducted a thorough characterization of a technology's nominal and tunable performance:

1. Math models that express the ideal function of the technology:

   a. First principles math models

      i. Linear approximation of $y = f(x)$

      ii. Nonlinear approximation of $y = f(x)$

   b. Empirical math models

      i.   Linear approximation of $y = f(x)$

      ii. Interactions between control factors

      iii. Nonlinear approximation of $y = f(x)$

2. Graphical models

   a. Geometric/spatial layout of the subsystem and platform technologies

   b. Component drawings in enough detail to build prototypes

3. Physical models

   a. Prototype subsystems

   b. Prototype components

   c. Measurement systems

      i. Transducers

      ii. Meters

    iii. Computer-aided data-acquisition boards and wiring

    iv. Calibration hardware, firmware, and software

4. Critical parameter-management data

    a. Mean values of all critical functional responses

    b. Baseline standard deviations of all critical functional responses (without the effects of noise factors)

    c. Cp of all critical functional responses (without the effects of noise factors)

    d. Coefficient of variation (COV) of all critical functional responses (without the effects of noise factors)

    e. List of control factors that have a statistically significant effect on the mean of all critical functional responses, including the magnitude and directionality of this effect

    f. List of control factors that have a statistically significant effect on the standard deviation of all critical functional responses, including the magnitude and directionality of this effect (without the effects of noise factors)

    g. List of control factors that have a statistically significant effect on the coefficient of variation (COV) of all critical functional responses (without the effects of noise factors)

    h. List of control factors that have statistically significant interactions with one another (including the magnitude and directionality of this effect)

    i. List of CFRs that have statistically significant interactions (correlations) with one another at the subsystem and system levels (including the magnitude and directionality of this effect)

We must be able to answer a simple question: "If I change this variable, what effect will that have on all the critical functional responses across the subsystem?" We have to get this "nominal performance knowledge" first, in the absence of statistically significant noise factors. We refine this knowledge in Phase 3, where the major emphasis is on intentional stress testing of the baseline subsystem technology with statistically significant noise factors. But before we can optimize robustness of the subsystem technology, we have to characterize its control factors for their nominal contributions to "Y is a function of x" relationships.

We must be able to account for critical correlations, sensitivities, and relationships within each new subsystem technology under nominal conditions. If we lack knowledge of these fundamental, critical relationships, the technology is risky due to our partial characterization of its basic functional performance. The development team has no business going forward until these things are in hand in a summary format for key business leaders and managers to assess. If they don't assess this evidence, they are putting the future cycle-time of the product commercialization projects at risk.

Gate 2 Readiness

If technology is allowed to escape Phase 2 with anemic characterization, it is almost a sure bet that someone down the line will have to pay for the sins of omission committed here. In a very real sense, the capital being invested in the new technology is being suboptimized. A great deal more capital will have to be spent to make up for the incomplete investment that should have occurred right here. You can either pay the right amount now (yes, it always seems a high price to pay) and get the right information to lower your risks, or be "penny wise and pound foolish" and really pay big later, when you can least afford it in both time and money. Phase 2 of technology development is truly the place to rigorously practice the old adage, "Do it right the first time."

A General List of Phase 2 Tools, Methods, and Best Practices

- Pugh concept-selection process
- Knowledge-based engineering methods

- Math modeling (functional engineering analysis)
- Concept-generation techniques (TRIZ, brainstorming, brain writing, and so on)
- Measurement system development and analysis
- Hypothesis formation and testing
- Descriptive and inferential statistics
- Statistical process control
- Capability studies
- Design of experiments
- Regression analysis
- Response surface methods
- Technology FMEA, fault tree analysis
- Modular and platform design
- System-architecting methods
- Prototyping methods (soft tooling, stereo lithography, early supplier involvement, and so on)
- Cost-estimating methods
- Risk analysis
- Critical parameter management
- Project-management methods
- Monte Carlo simulation
- Cycle-time FMEA

Table 5.2 integrates the Develop phase requirements, deliverables, tasks, and enabling tools.

TABLE 5.2 Tools, Tasks, Deliverables–to–Requirements Integration Table

Requirement	Deliverable(s)	Task(s)	Tool(s)
Superior technology concepts	Documented superior technology and measurement system concepts	Generate and refine technology concept evaluation criteria. Evaluate and select superior technology concepts.	Pugh concept-selection process Concept-generation techniques (TRIZ, brainstorming, brain writing, and so on)
Math models, prototypes, and measurement systems	Documented underlying math models (deterministic and statistical) that characterize the nominal and tunable performance (baseline) of the new technology concepts and their measurement systems Prototype hardware, firmware, and software that provides a physical model of the new technology subsystem concepts and their measurement systems	Analyze, characterize, model, and stabilize nominal and tunable performance of superior technology concepts (models, prototypes, and measurement systems).	Knowledge-based engineering methods Math modeling (functional engineering analysis) Measurement system development and analysis Hypothesis formation and testing Descriptive and inferential statistics Design of experiments Regression analysis Response surface methods Modular and platform design System-architecting methods Prototyping methods (soft tooling, stereo lithography, early supplier involvement, and so on)

Baseline functional stability, tunability, and capability under nominal conditions	Documented SPC charts (stability and tunability);capability indexes for all critical functional responses	Conduct SPC and capability studies under nominal conditions.	Statistical process control Capability studies Critical parameter management
Technology risk profiles	Documented technology risk assessment	Generate risk assessment and summary profile.	Technology FMEA Fault tree analysis Cost-estimating methods Risk analysis
Optimize phase project plans	Documented optimization project-management plans	Generate Optimize phase project plans and cycle-time model.	Project-management methods Monte Carlo simulation Cycle-time FMEA

I²DOV Phase 3: Optimization of the Robustness of the Baseline Technologies

Research and Technology Development

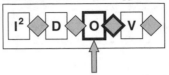

The Optimize gate requirements include the following:

- Robust and tunable technologies

- Stability, tunability, and capability under stressful conditions

- Technology risk profiles

- Verify phase project plans

The Optimize gate deliverables include these:

- Documented robustness of all technologies

- Documented stability, tunability, and capability of technologies under stressful conditions

- Updated documentation of critical parameter database

- Documented technology risk assessments

- Documented Verify phase project plans

The Optimize tasks include these:

1. Develop subsystem noise diagrams and platform noise maps.

2. Conduct noise factor experiments.

3. Define compounded noises for robustness experiments.

4. Define engineering control factors for robustness development experiments.

5. Conduct design for additivity analysis and run designed experiments to explore interactivity between engineering control parameters and noise parameters.

6. Analyze data and build a predictive additive model.

7. Run verification experiments to certify robustness gains from CF × NF interactions.

8. Document critical functional parameter nominal set points and CFR relationships.

9. Generate risk assessment and summary profile.

10. Generate Verify phase project plan and cycle-time model.

With the approval to enter the third phase of technology development, the new technologies have been proven to be worth the further investment required to optimize their performance. Optimization of a new technology focuses on two distinct activities:

- **Robustness optimization** of the Critical Functional Responses against a generic set of noise factors that the design will likely see in a downstream product commercialization application (per the deployment schedule based on the family plan strategy).

- **Certification of adjustment factors** that have the capability to place the mean of the technology's CFRs on to a desired target within the dynamic range specified by the application requirements from the Family Plan. Cpk can be made to equal Cp using these critical adjustment parameters.

We now have stable, tunable, and robust functions designed into our new technologies. Life will be simpler for commercialization teams that receive this kind of "safe" and "mature" technology. If you have ever been the recipient of overly sensitive, underdeveloped technology, you know exactly what value this can provide to a commercialization team.

From Phase 2, the superior technology concepts have been converted into prototype form and are proven to possess stable, tunable, functional behavior under nominal conditions within the comparative context of both analytical and empirical models.

The nominally characterized, conceptually superior technology prototypes are now refined to have the capacity to be evaluated as robustness test fixtures. Robustness test fixtures are made up of adjustable or easily changeable engineering parameters (commonly called control factors). These prototype fixtures are used to evaluate critical functional responses at various levels of control and noise factor set points. Their purpose is to investigate and measure interactions between control and noise factors. They are heavily instrumented to measure both control and noise factor set points (inputs) and critical functional responses (outputs). Noise factors can be changed during this phase of technology development. The goal in evaluating such fixtures is to identify control factor nominal set points that leave critical functional responses minimally sensitive to the debilitating effects of compounded noise factors.

The superior concepts are certified "safe" technologies into which additional resources can be invested to make them insensitive to commercial noise factors. Many R&TD organizations fall short of thorough noise testing by just evaluating new technologies in the presence of lab noise. Lab noise is not anything like use noise. Lab noise is usually random and not stressful. Use noise is many times stronger than lab noise and is not random. This is yet another source of risk that promotes lengthy schedule slips when partially robust and untunable technology is delivered into a commercialization project.

Attempting to make poorly-behaved and marginally-characterized technology robust later in product design can be a waste of time and money. This is why Taguchi and others recommend the development of dynamic technology based on well-characterized ideal functions, as was discussed in Phase 2. The certified math models represent the ideal functions to which Taguchi so frequently refers. As a rule, robustness-optimization activities should not come before the development of some form of credible analytical model. How rigorous and detailed the model should be is a matter of healthy debate that is based on economics as well as the application context. The rule is, don't waste your time trying to make a design that is unstable under nominal conditions robust under stressful conditions. Only stable, well-characterized designs are worth taking into the robustness-optimization process.

Review and Finalize CFRs for the New Subsystem Technologies

Before starting any optimization activities, the development team must have a well-defined set of critical functional responses. In addition, each CFR must have a measurement system defined with its capability documented in terms of repeatability and reproducibility (known as an R&R study) and precision-to-tolerance ratio (known as a P/T ratio). To produce viable results, continuous variables that are scalar or vector measures of functions that directly fulfill requirements must be used. We measure and make CFRs "robust" when quantifying the effects of control factor and noise factor interactions. We prefer not to count defects, but rather to measure functions within and across the integrated subsystems being developed as new platform technologies. This ensures that we measure "fundamental" Y outputs as we change "fundamental" x inputs.

During optimization, we assess the efficiency of the transformation, flow, or change of state of mass, energy, and information (logic and control signals that provide information about the flow, transformation, or state of the mass and energy of the subsystems). The efficiency is measured under the conditions of intentionally induced changes to both control and noise factors. To measure the flow, transformation, and state of these physical parameters, we have to move beyond a counting mentality to a measurement context. Although counting defects and assessing time between failures are measures of "quality," they are reactive metrics that are useful only after the function has deteriorated. We need to measure impending failure. This requires measuring functions in the presence of the noise factors that cause them to deteriorate over a continuum of time. We simply cannot wait until a failure occurs. This is a preventative strategy that leads us to our next topic: describing the activities that focus the development team's attention on the "physics of noise" that disrupt the flow, transformation, or state of mass, energy, and information. *This* is the science of reliability development. We value reliability prediction and evaluation, but they have little to do with the actual development of reliability.

Develop Subsystem Noise Diagrams and Platform Noise Maps

Noise has been described as any source of variation that disrupts the flow, transformation, or state of mass, energy, and information. More specifically, noise factors break down into three distinct categories:

- External sources of variation

- Unit-to-unit sources of variation

- Deterioration sources of variation

It is extremely useful to create a noise diagram (see Figure 5.4) that documents specific noise factors that fall under each of these categories.

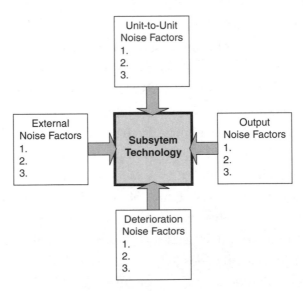

FIGURE 5.4   Subsystem noise diagram.

The cross-functional development team, as well as downstream advisors from design, production, service/support organizations, and selected customers must identify all the noise factors that potentially have a statistically significant effect on each of the subsystem critical functional responses. When a subsystem has a noise diagram constructed for each CFR, linkage can be established between each of the subsystem noise diagrams. Noise-transmission paths must be mapped within and between subsystem technologies. When two or more subsystem technologies are integrated to form a platform or some sort of higher-level system, we must characterize how noise from one subsystem affects the others. We also need to characterize the flow of noise from the platform out to other subsystems that might connect or interact with it. You can think of an engine as a platform made up of various subsystems, with the platform itself assembling into a car or truck or whatever type of product is being developed. If noise transmits beyond the platform and can reach a customer, service engineer, or other external constituency, that is important; we must document that potential negative relationship. When we account for these paths of noise transmission, we begin to understand how to alter the control factors to help guide the CFRs into a state of robustness against the noise. Insensitivity to noise is another way of communicating the meaning of robustness. If we don't have a comprehensive, system-wide noise map (with subsystem noise diagrams), how can we possibly begin the task of making a new subsystem technology or platform robust and know when we have finished?

Conduct Noise Factor Experiments

We just mentioned that subsystem noise diagramming and system noise mapping document noise factors that *might be* statistically significant. We can't tell whether any one of the noise factors is actually statistically significant (has an effect on the CFR that is statistically distinguishable from random events within a given confidence interval in the functional environment of the subsystem technology or platform) by sitting around a table talking about them in a conference room. It is imperative that the development team design an experiment to quantify the magnitude and directional effect of those noise factors selected for experimental evaluation. Failure modes and effects analysis can help identify and screen noise factors for inclusion in the noise experiment.

A noise experiment is usually a two-level fractional factorial designed experiment. It can be set up as a full or fractional factorial design. If strong interactivity is suspected, a minimum of a Resolution V fractional factorial design is recommended. If moderate-to-low interactivity is anticipated between the noise factors, a Resolution III Plackett-Burman screening experiment (an L12 array) is capable of providing a good set of data upon which to conduct an analysis of variance (ANOVA). ANOVA results provide the information required to tell whether any of the noise factors are statistically significant.

The subsystem prototype is set up at its baseline set point conditions and held there while the noise factors are exercised at their high and low set point conditions. The mean values of the CFRs are calculated. Analysis of means provides the data to explain the magnitude and directionality of each noise factor on the CFR.

Define Compounded Noises for Robustness Experiments

The results from a noise experiment provide the evidence required to go forward in defining how to properly and efficiently stress-test any given subsystem technology for robustness optimization. The statistically significant noise factors now can be isolated from the rest of the noise factors; they are the only noises worth using during robustness-optimization experiments. The others simply do not produce a significant effect on the mean value of the CFRs. Using the analysis of means data and its main effects plots, the development team can literally see the directional effect the noise factor set points have on the mean values of the CFRs. They can quantitatively demonstrate the effect of the significant noise factors—one more piece of guesswork is removed from the technology-development process.

Noise factor set points that cause the mean of a subsystem CFR to *rise* (significantly) are grouped or compounded together to form a compound noise factor, referred to as N1. Noise factor set points that cause the mean of a subsystem CFR to *fall* significantly are grouped or compounded together to form a compound noise factor, referred to as N2. Both compounded noise factors (N1 and N2) have the same set of statistically significant noise factors within them. The difference is the specific set points (high or low) that are associated with noise factor alignment with N1 or N2.

This noise-compounding strategy facilitates the unique and efficient gathering of two data points (data taken for each experimental treatment combination of control factors at N1 and N2 compounding levels) during the exploration of interactivity between control factors and noise factors. We need to be sure we have *real* noise factors identified because we want to perturb interactivity between control factors and noise factors to locate *robust* control factor set points. If the noises are not truly significant, we will not be exposing the control factors to anything that will promote the isolation of a true robust set point. Much time can be wasted testing control factors against noises that are little more than random sources of variation that do not represent what is really going on in the actual use environment. Many R&TD organizations evaluate only *nominal optimal performance* in the sterile and unrealistic environment of random and insignificant levels of noise. Their technology is then transferred to design in a state of relative immaturity and sensitivity. The design teams are left to deal with the residual problems of incapability of performance (this is certainly a design defect!). The usual response is, "That's funny, it worked fine in our labs ...." For those on the design team, it is anything but funny.

Define Engineering Control Factors for
Robustness Development Experiment

We now turn our attention to the control factors that will be candidates for changes to their baseline set points to new levels that leave the CFRs in a state of reduced sensitivity to the significant, compounded noise factors (N1 and N2). Selecting these control factors (see Figure 5.5) is not trivial. Not all control factors will help induce robustness—in fact, only those control factors that possess interactions with the noise factors will reduce sensitivity to noise.

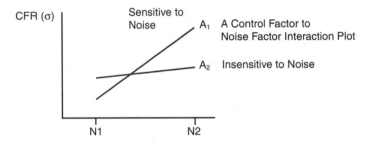

FIGURE 5.5   Robustness plot.

We begin control factor selection using the information about the ideal functions $(Y = f(x))$ gained during Phase 2. Any control factor that had a significant effect in the underlying functional model is a candidate control factor for robustness improvement. Actually, we will be looking for three kinds of performance from the candidate control factors:

- Control factors that have moderate to strong interactions with the compounded noise factors are good robustizing factors (good for minimizing the standard deviation, even with the noises freely acting in their uncontrolled fashion in the use environment).

- Control factors that have relatively weak interactions with the compounded noise factors *but* have strong effects on shifting the mean value of a CFR are critical adjustment parameters

(CAPs are good for recentering the function on a desired target, thus forcing Cpk = Cp).

- Control factors that have low interactivity with the compounded noise factors and low shifting affect on the mean of the CFR are good economic factors (allowing you to set them at their lowest-cost set points).

Design for Additivity and Run Designed Experiment

Using the knowledge from Phase 2 function modeling, assess the nature of the interactions between the control factors (CF × CF interactions). Apply the rules for additivity grouping or sliding levels based on how energy and mass are used in the ideal function to minimize or eliminate the disruptive effects of antisynergistic interactions. The experiment must be set up to focus on producing data that exhibits and highlights the interactivity between the control factors and the compounded noise factors. Within these CF × NF interactions, the robustness-optimization potential is found.

An appropriately-sized array is selected to evaluate the control factor main effects and any control factor × control factor interactions (usually a few two-way interactions that were statistically significant in the Phase 2 analysis). The typical array has a three-level fractional factorial format. If additivity grouping or sliding levels are used to suppress or remove the effects of unuseful CF × CF interactions, a Resolution III design (typically an L18 Plackett-Burman experimental array) commonly is used. If useful two-way interactions between control factors will be evaluated, a Resolution 5 or Full Factorial design must be used (remember, a Resolution IV design will likely confound the two-way interactions with one another). Because the compound noises are exposed equally to each row from the fractional factorial designed experiment, the control factor treatment combinations are aligned in a full factorial exposure to the noise. When you have "full" noise factor exposure to the fractional factorial control factor combinations, it ensures that the noise was faithfully and fully delivered to stress (interact with) the fraction of the possible design set points you selected. Using a fractional factorial noise-exposure strategy would understress the control factors and leave the design at risk of missing a CF × NF interaction that could be exploited on behalf of your customer's satisfaction! Those who think Taguchi Methods are optional or not effective might want to rethink that position; you are probably ignoring a whole class of interactions (CF × NF) that can make a big difference for your customers.

Analyze Data, Build Predictive Additive Model

The robust design experiment produces a rich set of data that two common statistical methods can analyze:

1. Analysis of means (ANOM)

   a. Provides insight on which level of each CF is best for robustness (largest S/N) or variance minimization (smallest σ or σ^2) of the CFR

   b. Provides insight on which control factors have the largest effect on the mean value of the CFR (critical adjustment parameters)

2. Analysis of variance (ANOVA)

   a. Provides insight on which control factors have the largest effect relative to one another and to the error variance in the entire data set (optional for S/N metric but quite useful for variance and mean analysis)

The development team uses the control factor set points that produced maximum improvements in CFR robustness to construct a unique math model, referred to as Taguchi's Additive Model. This model is used to predict overall robustness gain in units of the decibel. The additive model is a prediction of the expected robustness for the CFR under evaluation.

Run Verification Experiments to Certify Robustness

The "optimized" control factor set points must be verified in the final step of robust design. The same set of compounded noise factors is used during the verification experiments. At least five replicate verification tests commonly are run to prove that the predicted S/N, mean, and variance values are reproducible at these new set points. Baseline set point verification tests are also commonly conducted to see if the design acts the same way that it did back in Phase 2 testing, before robustness optimization. This helps ensure that the development team is not missing some additional control or noise factors that are important to the safe transfer of the technology for commercial application.

It is also common to verify the dynamic range of the critical adjustment parameters that affect the adjustment of the mean value of the CFRs. The best approach is to conduct further optimization of these CAPs by using regression or response surface methods.

Document Critical Functional Parameter Nominal Set Points and CFR Relationships

When the additive model and dynamic tuning capability of the subsystem technology have been verified, the new set points and tuning range limits must be documented. At this point, the critical parameter management database is updated with the results of the Phase 3 quantitative deliverables. The traceability of critical parameter relationships is greatly enhanced with the data from Phase 3.

Gate 3 Readiness

Phase 3 in the technology-development process is used to take technology that is safe (stable and tunable) into a state of maturity (insensitive to noise). Gate 3 is the place for assessing the proof that the new technology is robust to a general set of noise factors that you will likely encounter in commercial application environments. This is also the place to assess the tunability of the CFR to global optimum targets using one or more critical adjustment parameters.

A General List of Phase 3 Tools and Best Practices

- Subsystem noise diagramming
- System noise mapping
- Measurement system analysis
- Noise experiments
- Taguchi methods for robust design
- Design of experiments
- Analysis of means
- Analysis of variance
- Regression
- Response surface methods
- Statistical process control
- Capability studies
- Technology FMEA
- Critical parameter management
- Project-management methods
- Monte Carlo simulation
- Project FMEA

Table 5.4 integrates the Optimize phase requirements, deliverables, tasks, and enabling tools.

TABLE 5.4 Tools, Tasks, Deliverables–to–Requirements Integration Table

Requirement	Deliverable(s)	Task(s)	Tool(s)
Robust and tunable technologies	Documented robustness of all technologies	Develop subsystem noise diagrams and platform noise maps.	Subsystem noise diagramming
		Conduct noise factor experiments.	System noise mapping
		Define compounded ncises for robustness experiments.	Measurement system analysis
		Define engineering control factors for robustness development experiments.	Noise experiments
		Conduct design for additivity analysis and run designed experiments to explore interactivity between engineering control parameters and noise parameters.	Taguchi methods for robust design
		Analyze data and build predictive additive model.	Design of experiments
		Run verification experiments to certify robustness gains from CF × NF interactions.	Analysis of means
			Analysis of variance
			Regression
			Response surface methods

continues

TABLE 5.4 Tools, Tasks, Deliverables–to–Requirements Integration Table (continued)

Requirement	Deliverable(s)	Task(s)	Tool(s)
Stability, tunability, and capability under stressful conditions	Updated documentation of critical parameter database Documented stability, tunability, and capability of technologies under stressful conditions	Document critical functional parameter nominal set points and CFR relationships under stressful conditions. Conduct SPC and capability studies.	Statistical process control Capability studies
Technology risk profiles	Documented technology risk assessments	Generate risk assessment and summary profile. Conduct technology FMEA.	Technology FMEA
Verify phase project plans	Documented Verify phase project plans	Generate Verify phase project plan and cycle-time model.	Project-management methods Monte Carlo simulation Project FMEA

I^2DOV *Phase 4: Verification of the Platform or Sublevel Technologies*

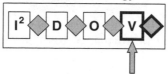

The Verify gate requirements include the following:

- Technology integration capability

- Technology that is capable under nominal and stressful conditions

- Reliability assessment

- Critical parameter management

- Technology risk profiles

- Technology transfer control plan

The Verify gate deliverables include these:

- Documented integration capability assessments

- Documented capability assessments under nominal and stressful conditions

- Documented reliability assessments

- Documented critical parameter-management databases

- Documented technology risk profiles

- Documented technology transfer control plans

The Verify tasks include these:

1. Integrate subsystems into platform or system integration test fixtures.

2. Refine and confirm "between and within" subsystem CFRs.

3. Develop, integrate, and verify platform/system integration data-acquisition systems.

4. Design platform or system integration nominal performance tests.

5. Conduct nominal performance tests (integration Cp verification).

6. Evaluate nominal performance data against technology requirements.

7. Design platform or system integration stress tests (integration Cpk verification).

8. Conduct platform or system-integration stress tests.

9. Evaluate stressed performance data against technology requirements.

10. Perform ANOVA on stress data to identify sensitivities to integration noises.

11. Refine, balance, and improve platform or integrated subsystem performance.

12. Rerun stress tests and perform ANOVA on data to verify reduced sensitivities.

13. Conduct preliminary reliability models and evaluations for new subsystem technologies and platforms.

14. Document robustness, tuning, and CFR relationships, and certify that the technology is transferable.

15. Conduct final technology risk assessment.

16. Prepare technology transfer control plan.

The final phase of technology development primarily focuses on integrating *and verifying* the newly robust and tunable subsystem technologies into an existing product architecture or into a new platform architecture. The development of subsystems and platforms is certified for commercial readiness at the Phase 4 gate. The transfer of the new technologies to product design depends on transferability criteria (technology requirements) that will be fulfilled in this final phase of technology development.

The evaluations that are conducted during this phase are stressful by nature, to help ensure the technologies are mature enough to transfer. Some technologists are averse to stress testing their "pampered" designs because they fear that they will be rejected and will not be used in the next new product. When these individuals successfully protect their designs, they place every other employee in the company at risk and do great disservice to the customers of the business. Most technology problems show up in product design when technologies have been prematurely deployed in a commercialization project. They were not thoroughly stress-tested for robustness (S/N and σ) and capability (mean, Cp, and Cpk). They look and work fine until they are used in a context in which the real noise factors are applied to them (intentionally or otherwise). At that point, their true character emerges and all hell breaks loose! They were not completely characterized (usually to save face, time, and money), so they were transferred at great risk. When we reactively use the checklists and scorecards at the end of each of the four gates of technology development, we can expect to *routinely* uncover the risk that is building from anemic and incomplete characterization. Risk can be managed much more efficiently if one proactively uses the checklists and scorecards to preplan for the successful application of a balanced set of tools and best practices. If we choose to bypass certain tools and best practices, the technology will accrue certain risks. The gates are designed to assess the accrual of risk and to stop or alter development when the risk gets too high.

The fourth, "integrative" phase is extremely important and the one that most companies bypass with reckless abandon. When development teams fail to integrate their new subsystem technologies into an existing product or the new platform and stress-test them with statistically significant noise factors, they forego obtaining the knowledge that ensures that the risk of transferring the new technology is acceptable. Many say, "We can't afford to develop technology in this rigorous fashion." They go on to claim, "We like to integrate and test (usually with very low levels of statistically insignificant noise) early during product design and find out where our problems are and then fix (react to) them." This is called "build-test-fix" engineering and is antithetical to preventing functional performance problems. This approach ensures that there *are* problems and seeks comfort in isolating and "fixing" them—typically over and over again. This is a recipe for high warranty costs, late product launches, many rounds of costly post-shipping modification kits, aggravated and overworked service engineers, and the loss of once-loyal customers. Even worse, from a strategic perspective, is the fact that this approach ties up R&TD resources and design resources on lengthy "reliability emergency assignments" that keep them away from working on the next design project that is critical to future revenue generation. Build-test-fix engineering robs the corporation of its future generation of value, productivity, and economic growth. All the excuses for build-test-fix engineering end up being the most costly and inefficient way of doing business.

Integrate Subsystems into Platform or Product Test Fixtures

The optimized subsystems now can be integrated into one of two types of systems for stress testing and performance certification:

- An existing product

- A new platform that will form the basis of many new products

When the new subsystem or subsystems have been developed for integration into an existing product that is currently in the marketplace, the R&TD organization must obtain a sufficient number of these systems for stress testing and certification of integrated performance capability.

When the new subsystems are developed for integration into a new platform, a significantly different effort must be made to integrate the subsystems into a functioning system that we recognize as a platform. This is harder and requires more effort because there is no existing "product" to be reworked to accept the new subsystems. The platform is a whole new system. It is entirely possible that numerous existing subsystem designs will be reused and integrated into the new platform architecture during product design.

Refine and Confirm "Between and Within" SS CFRs

Before conducting certification tests, the development team needs to identify exactly how the critical functional responses will be measured as they transmit energy, mass, and information (analog and digital logic and control signals) across subsystem boundaries. Every subsystem already will have an internal or within set of CFRs defined for it. These were developed and measured in the first three phases of technology development. Now that the team is going to integrate all the new subsystem technologies together, it is highly likely that some need will arise to measure how the CFR from one subsystem affects the mating subsystem. If one subsystem receives mass from another subsystem, the mass flow rate will need to be measured as a "between" subsystem CFR. If the mechanism for mass flow or transfer is through the generation of a magnetic flux, that is classified as a "within" subsystem CFR. The important issue here is for the team to be comprehensive in accounting for CFRs that measure functional interface relationships. One helpful way to approach this accounting of CFRs at integration is to write energy-balancing equations to account for the flow, transfer, or state of change between energy and mass across the interface boundary. These "between" subsystem CFRs might have been defined previously during Phases 2 and 3 but were not measured in the real context of an integration test. The task now becomes one of measurement system development and analysis, our next topic.

Develop, Integrate, and Certify Platform Data-Acquisition Systems

A data-acquisition system capable of measuring all "within and between" subsystem CFRs, as well as platform- or product-level CFRs, needs to be developed, integrated, and certified at this point. These systems include numerous transducers that physically measure the CFRs, the wiring, interface circuit boards, the computers and software that facilitate the collection, processing, and storage of the CFR data.

The data-acquisition system needs to have its repeatability, reproducibility, and precision-to-tolerance ratio evaluated to ensure that the measurements are providing the level of precision and accuracy required to certify the performance of all the CFRs.

Design Platform or Integrated SS Nominal Performance Tests

The team must structure a series of performance tests to evaluate the Cp and Cpk of the CFRs under nominal conditions. This test is designed to examine the performance of all the CFRs under non-stressful conditions. It represents the best you can expect from the subsystems and the integrated platform or product at this point in time. These tests usually include normal use conditions and scenarios. All critical-to-function parameters within the subsystem and critical functional specifications at the component level are set at their nominal values. Environmental conditions are also set at nominal or average conditions.

Conduct Nominal Performance Tests

With a comprehensive nominal condition test plan in hand, the team can move forward and conduct the nominal performance tests. The data-acquisition system is used to collect most of the performance data from the CFRs. Some independent, special forms of metrology likely could be needed to measure some of the CFRs, typically at the platform or product levels. These are often called *offline* evaluations. In some cases, data must be gathered on equipment in another lab or site equipped with special analysis equipment.

Evaluate Data Against Technology Requirements

The data now has been summarized and evaluated against the technology requirements. Here we discover just how well the CFRs are performing against their targets, established back in Phase 1. This likely will be the best data the team will see because it was collected under the best or nominal conditions. With the nominal capabilities of the CFRs in hand, the team is ready to get an additional set of capability data that is run under the real and stressful conditions of noise.

Design Platform or Integrated Subsystem Stress Tests

Exposing new technologies to a statistically significant set of noise factors was discussed during the Phase 3 robustness-optimization process. We return to that topic within the new context of the integration of the robust subsystems. To be circumspect regarding realistically stressing the new platform technologies or existing products that are receiving new subsystem technologies, the team must expose these systems to a reasonable set of noise factors.

The noise factor list typically includes numerous sources of variation that are known to provoke or transmit variation within the system integration context. A designed experiment is the preferred inspection mechanism that is capable of intentionally provoking specific levels of stress within the integration tests. Designed experiments help ensure that thorough and properly controlled amounts of realistic sources of noise are intentionally introduced to the integrated subsystems. Without this form of designed evaluation of the various CFRs associated with the integrated subsystems, the technology is not fully developed and cannot be safely transferred into product commercialization.

Conduct Stress Tests

Conducting a stress test under designed experimental conditions requires a good deal of planning, careful setup procedures, and rigorous execution of the experimental data-acquisition process. As is always the case when measuring CFRs, the data-acquisition equipment must be in excellent condition, with all instruments within calibration and measurement system analysis standards.

Perform ANOVA on Data to Identify Sensitivities

The data set from the designed experiment can be analyzed using analysis of variance methods. ANOVA highlights the statistical significance of the effects of the noise factors on the various CFRs that have been measured. It is extremely important that the team be able to document the sensitivities of each CFR to specific integration noise factors. This information is extremely valuable to the product design engineers who will obtain the transferred technologies. If the integrated system of subsystems is too sensitive to the induced noise factors, it could be necessary to return the technologies to Phases 2 and 3, where the subsystems can be changed and recharacterized for added capability and robustness.

Refine and Improve Platform or Integrated Subsystem Performance

If the integrated subsystems are shown to have sensitivities that can be detuned without returning the subsystems to Phases 2 and 3, that should be done at this point. It is not unusual to find that subsystems need to be adjusted to compensate for interface sensitivities and noise transmission. Sometimes a noise factor can be removed or its effect can be minimized through some form of noise suppression or compensation. Typically, it is possible, but expensive, to control some noise factors. A common form of noise compensation is the development of feedback or feed-forward control systems. These add complexity and cost to the technology, but sometimes this is unavoidable.

Rerun Stress Tests and Perform ANOVA on Data to Verify Reduced Sensitivities

After changes have been made to the subsystems and noise-compensation strategies have been deployed, the subsystems can be integrated again and a new round of stress testing can be conducted. The ANOVA process is conducted again and new sensitivity values can be evaluated and documented.

Conduct Reliability Evaluations for New Subsystems Technologies and Platforms

Each subsystem technology and integrated platform has its reliability performance evaluated at this point.

Document CFPs and CFR Relationships and Certify That the Technology Is Transferable

The ANOVA data also holds capability (Cp) information for each CFR. This data is in the form of variances that can be derived from the mean square values within the ANOVA table (divide the total mean square by the total degrees of freedom for the experiment to obtain the total variance for the data set). If certain noises are applied to carefully selected critical functional parameters (usually as either tolerance extremes or deteriorated functional or geometric characteristics), the sensitivities of the changes at the critical functional parameter level (x inputs) can be directly traced to variation in the CFRs (Y outputs). This is an important final step in critical parameter characterization within the final phase of technology development.

Gate 4 Readiness

This last phase of technology development requires additional summary analysis and information related to the following topics:

- Patent stance on all new technologies
- Cost estimates for the new technologies

- Reliability performance estimates and forecasts

- Risk summary in terms of functional performance, competitive position, manufacturability, serviceability, and regulatory, safety, ergonomic, and environmental issues

Every company is different, so your specific areas of summary for technology development information transfer might contain items beyond what is listed here.

A General List of Phase 4 Tools and Best Practices

- Nominal performance-testing methods

- Measurement system analysis

- Stress-testing methods using designed experiments

- ANOVA data analysis

- Empirical tolerance design methods

- Sensitivity analysis

- Reliability modeling and estimation methods

- Monte Carlo simulation

- Cost-estimation methods

- Patent search and analysis

- Robustness evaluation

- Capability studies and characterization

- Technology transfer control plan methods

Table 5.5 integrates the Verify phase requirements, deliverables, tasks, and enabling tools.

TABLE 5.5 Tools, Tasks, Deliverables-to-Requirement Integration Table

Requirement	Deliverable(s)	Task(s)	Tool(s)
Technology integration capability Technology that is capable under nominal and stressful conditions	Documented integration capability assessments Documented capability assessments under nominal and stressful conditions	Integrate subsystems into platform or system integration test fixtures. Refine and confirm "between and within" subsystem CFRs. Develop, integrate, and verify platform/system integration data-acquisition systems. Design platform or system integration nominal performance tests. Conduct nominal performance tests (integration Cp verification). Evaluate nominal performance data against technology requirements. Design platform or system integration stress tests (integration Cpk verification). Conduct platform or system integration stress tests. Evaluate stressed performance data against technology requirements. Perform ANOVA on stress data to identify sensitivities to integration noises. Refine, balance, and improve platform or integrated subsystem performance. Rerun stress tests and perform ANOVA on data to verify reduced sensitivities.	Nominal performance-testing methods Measurement system analysis Stress-testing methods using designed experiments ANOVA data analysis Empirical tolerance design methods Sensitivity analysis

TABLE 5.5   Tools, Tasks, Deliverables–to–Requirement Integration Table (continued)

Requirement	Deliverable(s)	Task(s)	Tool(s)
Reliability assessment	Documented reliability assessments	Conduct preliminary reliability models and evaluations for new subsystem technologies and platforms. Conduct technology FMEAs.	Reliability modeling and estimation methods Monte Carlo simulation
Critical parameter management	Documented critical parameter management databases	Document robustness, tuning, and CFR relationships and certify that the technology is transferable.	Robustness evaluation Capability studies and characterization
Technology risk profiles	Documented technology risk profiles	Conduct final technology risk assessment.	Technology FMEA Cost-estimation methods Patent search and risk analysis
Technology transfer control plan	Documented technology transfer control plans	Prepare technology transfer control plans.	Control plan methods

References

Clausing, Don P. *Total Quality Development.* ASME Press, 1994.

Wheelwright, Steven C. *Revolutionizing Product Development.* Free Press, 1992.

Cooper, Robert G. *Winning at New Products: Accelerating the Process from Idea to Launch.* Perseus Books Group, 2001.

Christensen, Clayton M. *The Innovator's Dilemma.* Harvard Business School Press, 1997.

Utterback, James M. *Mastering the Dynamics of Innovation.* Harvard Business School Press, 1996.

Meyer, Marc H. and Alvin P. Lehnerd. *The Power of Product Platforms: Building Value and Cost Leadership.* Free Press, 1997.

TABLE 5.6 Gate Deliverable Review Scorecard

1	2	3	4	5	6
Gate Deliverable	Grand Avg. Tool Score	% Task Completion	Results vs. Requirement	Risk Color Code(R-Y-G)	Gate Requirement

TABLE 5.7 Task Scorecard

1	2	3	4	5	6
Gate Deliverable	Grand Avg. Tool Score	% Task Completion	Results vs. Requirement	Risk Color Code(R-Y-G)	Gate Requirement

TABLE 5.8 Tool Scorecard

1	2	3	4	5	6
Tool	Quality of Tool Use	Data Integrity	Results vs. Required Task Deliverable	Specific Tool Score	Task & Deliverable

6

TACTICAL PRODUCT COMMERCIALIZATION PROCESS

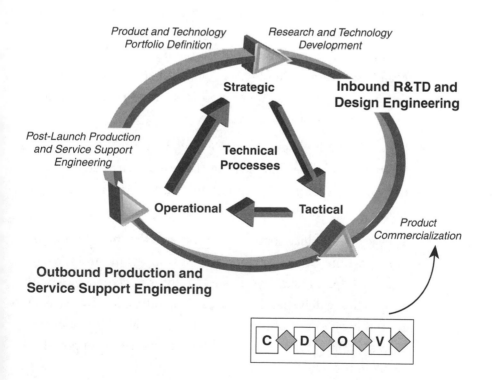

Six Sigma–Enhanced
Product Commercialization

The use of phases and gates to manage the timing and flow of product commercialization projects has shown strong results in many companies. Some skeptics have stated that such approaches limit creativity, constrain the early discovery of problems, and hold back the speedy launch of new products. We have found that companies with such spokesmen possess some common and repeatable traits:

- A single personality or a small clique of insiders claim to represent the voice of the customer. They possess a driving influence that is used to define new-product concepts and requirements in relatively ambiguous terms that slowly enable the development team to "back themselves into a corner" to a default system architecture and "final" product specifications relatively late in the product design timeline.

- A focus on the early, "speedy" integration of immature and underdeveloped subsystems that then require seemingly endless cycles of corrective action. Frequently, no value is placed upon a separate technology-development process or team to ensure the safety of technology as it is transferred into a fast-paced commercialization project.

- A loosely defined set of skills enabled by the undisciplined practice of a limited portfolio of tools and best practices that are centered on reacting to problems "discovered (actually, created)" in (see the second bullet in this list).

- A legacy of residual post-launch problems that tie up a significant portion of the company's development and design resources, precluding work on the future technologies and designs that will be needed to create the future revenue stream for the business. Problems were not prevented. The current process is fairly efficient at "problem assurance" as opposed to problem prevention.

- A "modification kit" stream that contains the true optimized and final designs that send a message to the customer that "you have been our guinea pig while we finish designing this product—after you have bought and paid for it." Recalls and warranty payouts erode the promised returns from the business case.

Most companies have embraced a phase-gate process, but deploy it in a *passive,* hands-off manner. They have a reasonable process—they just don't use it very well. Companies such as PRTM do a good job designing "management excellence frameworks" that suffer from two problems: They don't get used very well and the phase-gate process is void of task-tool alignment definition that enables functional excellence within the teams that actually do the work of product development.

Phase-gate-oriented companies that attempt to codify "management excellence" have made a small step forward but are still clinging to many of their past bad habits. Their legacy is similar to the one just covered: poor system performance and reliability, numerous subsystem-redesign cycles, a modification kit strategy to finish the design while it is in the customers' hands, cost over-runs with residual high cost of ownership, and missed market windows. What will fix this mess? In our view, a strong blend of management and functional excellence that ensures the linkage between what is required and what is actually done as a designed flow of DFSS-enhanced work.

This chapter develops a very active and disciplined approach to the use of DFSS-enhanced phases and gates during product commercialization. The goal is simple: to *prevent* the things found in the previous bulleted list.

Preparing for Product Commercialization

A few key things must be in place, *in this order,* before proper product commercialization can take place:

- Defined business goals and a strategy to meet them

- Markets and segments that will be used to fulfill the business goals

- Long-range voice of the customer and voice of technology defined based on specific market segments

- Product portfolio plan, product line strategy, and family plans defined to fulfill specific market segment needs

- Technology strategy defined to fulfill product line strategy

- New platform and subsystem technologies developed and certified within their own independent technology-development process (I²DOV—see Chapter 5, "Strategic Research and Technology Development Process")

With these things in place, it is much easier to see what product flow to commercialize and when each individual product from the family plan should be developed and delivered to the market (see Figure 6.1). When we develop and deliver a specific product or service to the market, we call this flow of work commercialization. Well-structured commercialization increases the probability of sustainable growth for the business and its shareholders. Design for Six Sigma helps structure tasks during commercialization and provides a portfolio of selectable tools, methods, and best practices that makes it easier to design "lean" workflows. DFSS adds value to top-line-growth financial results through its capability to enable and deliver results that fulfill the specific product requirements that underwrite the validity of the product's business case.

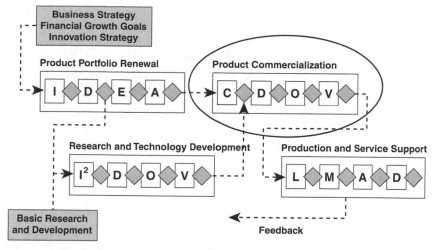

Figure 6.1 Integrated process model.

How do we value the use of Six Sigma in product commercialization? We do it through fulfilling the product's business case and avoiding the downstream cost of poor quality. In this context, the Taguchi loss function can be helpful in quantifying the cost avoidance induced by preventing design problems. Taguchi's loss function simply states that when a product's function deviates from a desired target, customers, those who deliver and service the product, and, in many cases, society at large suffer economic loss. Designing products that are robust to sources of destabilizing variation helps minimize these financial losses. When you hear the word *robust*, think of something that is hard to destabilize or knock off target, even under stressful conditions. Design for Six Sigma can also be valued through the direct cost savings realized by the redesign of products and production processes that are causing warranty, scrap, and rework costs. DFSS can be used on a reactionary basis to clean up costly design problems, but it is best applied proactively during commercialization to prevent problems.

In terms of delivering a complete, well-integrated system, the methods of systems engineering and integration must be in place as a governing center for the product commercialization project (see

Chapter 7 on systems engineering and integration in *DFSS for Technology & Product Development*; Prentice Hall, 2003). The whole project must be managed from the technical perspective of systems engineering, not from the parochial perspectives of a loosely networked group of subsystem or sublevel design managers. If you are sensing a form of management rigor that runs the project with a firm, disciplined hand from a project-management perspective, that is much preferred over the "bohemian-existential" approach of poorly linked subsystem teams. A systems-engineering team should be in place; otherwise, personal agendas and political behavior will run the show. One of the most common problems we have encountered in the world of commercialization is the lack of well-designed and well-run systems-engineering teams. If this continues, it's back to the five things we mentioned at the beginning of this chapter.

Defining a Generic Product Commercialization Process Using the CDOV Roadmap

We use a generic and simple model to illustrate a phase-gate structure to conduct product commercialization. The CDOV process phases we use in this text serve as guideposts to help you align the tools and best practices of Design for Six Sigma to the actual phases in your company's commercialization process. *CDOV* stands for the first letters in the four phases shown in Figure 6.2.

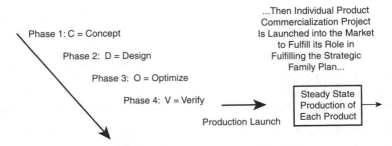

FIGURE 6.2   CDOV commercialization process model.

It is not uncommon to add subgates within each of the four phases. This is particularly true of Phases 3 and 4. Phase 3 often requires subgates because of the need to first develop robust subsystems and then to integrate them into a system for final performance balancing under nominal and stressful conditions. Phase 4 requires subgates to verify the final product design and then moves on to verify production ramp-up, launch, steady-state operations, supply chain, and support processes. We illustrate how they flow as part of the overarching DFSS-enhanced product commercialization process that leads to successful product launches.

The use of gates requires that the date for a gate be firmly established. As mentioned earlier, the criteria for gate passage is flexible and, in some cases, non-negotiable. Thus, we illustrate the nature of phase-gate transition as an overlapping interface. Some deliverables will be late, while others will come in early.

The program-management team can, at its discretion, allow some activities to start early or end late.

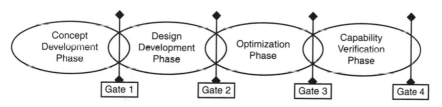

FIGURE 6.3   CDOV phase-gate overlap model.

We define the general nature of each of these phases and describe the detailed steps within each phase. Because this book focuses on the technical development of products in the context of Six Sigma tools, we do not spend a lot of time on business case development and some other details that must be included in a comprehensive product-development process. For the sake of being circumspect in making the reader aware of these additional issues, we briefly illustrate what a general phase and gate should include by topical area:

Phase topics:

- Business/financial case development

- Technical/design development

- Manufacturing/assembly/materials management/supply chain development

- Regulatory, health, safety, environmental, and legal development

- Shipping/service/sales/support development

We focus most of our discussions on the technical/design development and manufacturing/assembly/materials management/supply chain development areas. It is important to recognize, though, that a broad range of Six Sigma methods can be applied across all the phase topics listed here. That broader range includes Six Sigma methods for problem solving or process redesign in transactional business processes, manufacturing operations processes, and service processes—all of which deploy the DMAIC or DMADV roadmap.

The best way to think about establishing the architecture of any product-development process is to structure it as a macro timing diagram to constrain "what to do and when to do it" across the entire product development lifecycle. With such a macro timing diagram established, the product-development management team can conduct the phases and gates with the help of checklists and scorecards based on tasks enhanced by a broad portfolio of tools and best practices and their deliverables. The techniques of project management can then be used to construct micro timing diagrams (preferably in the form of a PERT chart) that establish discipline and structure within the day-to-day use of the tools and best practices to complete key tasks for each person on the product-development team. We recommend PERT charts because they help define personal accountability for completing tasks using the right tools and best practices within the phases (see Chapter 3, "Project Management in Technical Processes," for a full discussion). They also greatly enhance the management of parallel and serial activities, workaround strategies, and the critical-path tasks that control the cycle time of each phase. Managing a commercialization project in this manner helps ensure that all the right value-adding things get done to balance the ever-present cost, quality, and cycle-time requirements. This is what we call functional excellence.

No one product-development process architecture will meet the needs and priorities of all companies. Every company must define a product-development process that fits its culture, business process, and technological paradigm. Recall that CDOV is just our way to initiate communications about linking DFSS to your phases and gates; we fully expect you to use your own names for your phases.

Let's move on and review the details of the phases and gates of the CDOV process. As we do this, we begin to develop the critical parameter-management paradigm that is so helpful in development of Six Sigma performance.

The CDOV Process and Critical Parameter Management during the Phases and Gates of Product Commercialization

The following diagram illustrates the major elements for CDOV.

Product Commercialization

Concept Phase: Develop a System Concept Based on Market Segmentation, the Product Line, and Technology Strategies

The Concept gate requirements include the following:

- Product requirements derived from current voice-of-the-customer data

- Superior product concept derived from a variety of competitive alternatives

- Certified technologies from the R&TD process (IIDOV phases)

- Functional and reliability models for product/system

- Technology and market risk profiles

- Design phase project plans

The Concept gate deliverables include these:

- Documented voice-of-the-customer data (structured, ranked, and prioritized; KJ document)

- Documented new, unique, and difficult requirements (product House of Quality)

- Documented easy, common, and old requirements (general requirements document)

- Documented system functional models

- Documented system concept alternatives

- Documented final, superior system (product) concept

- Documented technology transfer control plans

- Documented risk profiles for our technology and markets

- Documented project plans for the projects moving into the Develop phase

The Concept tasks include these:

1. Gather the specific voice of the customer by market segment for the idea bounded by the project being activated from the portfolio.

2. Structure, rank, and prioritize the voice of the customer using KJ analysis.

3. Create product- or system-level House of Quality.

4. Conduct competitive product benchmarking.

5. Generate product- or system-level requirements document.

6. Define the functions that fulfill the system requirements.

7. Generate system-level concept-evaluation criteria.

8. Generate system-level concepts that fulfill the functions.

9. Evaluate system-level concepts.

10. Select superior system-level concept.

11. Analyze, characterize, model, and predict nominal performance of the superior system.

12. Develop reliability requirements, initial reliability model, and FMEA for the system.

13. Generate risk assessment and summary profile.

14. Generate Develop phase project plans and cycle-time models.

This section reviews the specific details related to each action item (the start date for each action item, the action item, and the end date for each action). Each of these action items can be conducted by using a tool or best practice. The flow of actions can be laid out in the form of a PERT chart for project-management purposes (see Chapter 4, "Strategic Product and Technology Portfolio Renewal Process").

Gather the Specific Voice of the Customer by Market Segment for the Idea Bounded by the Project Being Activated from the Portfolio

The best way to start any commercialization project is to know two key things well:

- **The voice of the customer**—What your customers need

- **The voice of technology**—What product and manufacturing technology is currently capable of providing

We lay out the importance of gathering detailed information regarding both of these foundational areas in the next few paragraphs. These two sources of information will form the basis of what the entire product commercialization community will design to fulfill or surpass. Critical parameter management is initiated by generating a database of ranked, prioritized, stated, and observed customer needs. It is also baselined with current technology parameters that can perform to known levels of capability.

It is helpful to begin product design activities by asking questions such as these: What do customers specifically say they need? What do their actions suggest they need? What technology is being used today to meet their current needs? What new technology (manufacturing and product design technology) is capable of extending or improving their satisfaction with a new product? Notice that all the actions implied in these questions focus on fulfilling stated or observed/ implied customer needs.

One of the biggest problems in starting the process of product commercialization is balancing the voice of the corporate marketing specialist with the voice of the customer. Most companies underinvest in gathering an up-to-date and thorough database of actual customer needs. It is much cheaper and easier to get a little data once in a while and then use intuition, experience, personal opinion, and well-intentioned best guesses to define a surrogate body of information that is submitted as representative of the voice of the customer. It has been our experience that many marketing groups have weak market-research processes, a poorly-developed set of tools and best practices, very limited budgets, and little cross-functional linkage to the engineering community. We have addressed this by writing the world's first comprehensive text on this in *Six Sigma for Marketing*

Processes (Prentice Hall, 2006). All this adds up to the most common form of VOC data being characterized by the following traits:

- The "data" is frequently old, often by 1 or more years.

- The marketing team is incapable of showing the actual words taken from a customer interview—"data" is usually documented in the memory of the market specialist or written down in the "voice" of the market specialist.

- The "data" is not structured and stored in a format that is easy to translate and convert into a ranked and prioritized set of product requirements.

- The information in hand is from a limited set of "key" customers that is not truly (or statistically) representative of the true market segment being targeted.

- The "data" is the "considered opinion" of a strong-willed and opinionated executive (sometimes it's an R&D or engineering VP, sometimes it's a marketing or sales executive, and sometimes it's the CEO).

- Engineers who are developing the product and manufacturing/service and support processes were not involved in gathering the voice-of-the-customer data (nor have they been heavily involved in competitive performance benchmarking).

The real problem here is that developing products and their supporting manufacturing processes in this anemic context places the business at great risk. The risk lies in the under- or overdevelopment of a product based on a poor foundation of incoming customer need information and competitive product design and manufacturing process technology. The business is essentially guessing or relying on intuition. Of course, if you have unlimited financial resources, this is fine. Developing hunches into products can be a tremendous waste of time and money. This is certainly no way to help ensure sustainable growth.

A major antidote to the improper development of product requirements is to go out into the customer use environment and carefully gather the real voice of the customer. Product commercialization teams can accomplish this by planning the use of a proven set of VOC gathering tools, methods, and best practices during the first phase of product commercialization.

Numerous methods are available to facilitate the gathering of customer needs data. By category, really only two general forms of data exist:

- Customer needs defined by *what they say*

- Customers needs defined by *what they do or try to do*

One of the most effective methods for gathering data about stated customer needs is to form a small two- to three-person team and hold a face-to-face meeting with a reasonable cross-section of customers. One person leads the discussion, one takes detailed notes or records the conversations, and one listens and adds to the discussion to help keep a diverse and dynamic dialogue in motion. This approach is characterized by a very active and dynamic exchange between the customers and the interview team, which we call the inbound marketing team. The inbound marketing team is typically a cross-functional group made up from a reasonable mix of professionals in the following areas:

- Market research and advanced product planning

- Technical (R&D, system or subsystem design engineering, or production operations, as appropriate)

- Industrial design

- Service and support

- Sales

- Program management

The inbound marketing team first prepares a customer interview guide that enables members to set a specific flow of discussion in motion during the customer meeting. The interview guide is not a questionnaire, but rather a logical flow diagram containing specific topics that are designed to stimulate free-flowing discussions about customer needs and issues to help align the advanced product-planning activities of the company with the future needs of the customer. If a questionnaire is all that your business needs to define future product-development requirements, then phone calls, hand-out questionnaires, and mailing methods are much more economical. Questionnaires are quite limited in the kind of detailed data that can be obtained. This because you get very little exchange and clarification of information to verify that what was stated was correctly received. The power of face-to-face discussions with a reasonable number of customers (current, former, and potential) is unsurpassed in ensuring that you obtain many key VOC attributes (based on the KANO model of customer satisfaction):

- Basic needs that must be in the future product

- Linear satisfiers that produce proportional satisfaction with their increased presence in the future product

- Delighters that the customer generally is not aware of until you stimulate some sense of latent need that was "hidden" in the customer's mind

The other key data that is available to the inbound marketing team is the observed behavior of customers in the product acquisition and use environment. These are relatively passive events for the inbound marketing team, but obviously dynamic events on the customer's side. Here the team watches and documents customer behavior instead of listening to what they say they need. This approach has been called *contextual inquiry* (see the book *Total Quality Development;* ASME Press, 1994). This method also requires careful preplanning so that meaningful vantage points and contexts for observation can be arranged. The team must find a way to participate in usage experiences. Obviously, not all products are amenable to this approach, but when the product is used in an observable environment, it can be quite useful in assisting in the determination of unspoken needs. In the world of KJ analysis (our next topic), this data is profoundly useful in developing images of customer problems and needs. A well-circulated story tells of Honda engineers positioning themselves in parking lots and watching how people used their trunks and hatchbacks, to help gather data for improving the next hatchback design. Eastman Kodak hands out one-time-use cameras to their employees and assesses their use habits. You can align your teams with use experiences in many creative ways. Never underestimate the knowledge you can gain by getting inside the situational events that are associated with the actual use of your products and processes—especially when disruptive sources of variation (causes of loss) are active within the use environment.

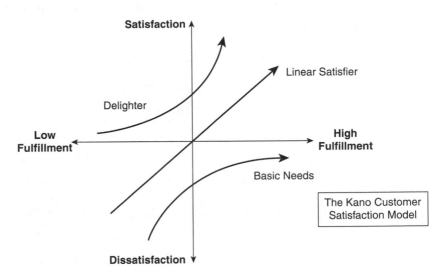

FIGURE 6.4   The Kano model.

Refine and Rank the Voice of the Customer Using KJ Analysis

After the customer data is gathered in sufficient quantity to adequately represent the scope of the market segment, the team refines and documents the raw data by structuring it into a more useful form. The KJ method, named after the Japanese anthropologist Jiro Kawakita, is used to process the VOC data into rational groups based on common themes (higher levels of categorizing the many details of expressed or observed/imaged need). The term *affinity diagramming* has been used to characterize this process.

KJ analysis is quite useful when you are working with non-numerical data in verbal or worded format. At first glance, the mass of VOC input can seem to be a jumble of multidimensional and multi-directional needs that are quite hard to get your hands around. In practical terms, KJ analysis helps make sense of all these words, observations, and images by aligning like statements in a hierarchical format. After the raw VOC is segmented, aligned underneath higher-level topical descriptors (higher levels of categorizing the many details of expresses or observed/imaged need), and ranked under these top-ical descriptors, it can be loaded into the system- or product-level House of Quality using quality function-deployment techniques.

Create Product- or System-Level House of Quality

A process known as quality function deployment is commonly used to help further refine the affinitized (grouped by similar characteristics) VOC data. QFD uses the data, mapped using the KJ method, to develop a clear, ranked set of product-development requirements that guide the engineering team as it seeks to determine what mea-sures of performance need to be at Six Sigma levels of quality.

The QFD process produces a two-dimensional matrix called a House of Quality. At the system level, the House of Quality is used to transform VOC data into technical and business performance requirements for the new product or process. Many teams overdo the structuring of the House of Quality by including too many minor details in it. The House of Quality should receive only VOC data that falls under one or more of the following categories:

- **New needs**—Features or performance that your product has not historically delivered

- **Unique needs**—Features or performance that are distinctive or highly desired beyond the numerous other less-demanding needs that must be provided

- **Difficult-to-fulfill needs**—Features or performance that are highly desired but are quite difficult for your business to develop and will require special efforts/investments of resources on your part

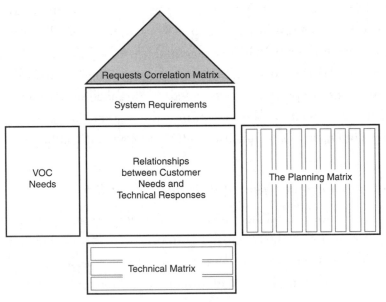

The House of Quality, as defined by Lou Cohen in *Quality Function Deployment* (Addison-Wesley, 1995)

FIGURE 6.5   The House of Quality.

Everything that will be included in the product design requirements document is important—but not everything is critical. The House of Quality is a tool for critical parameter identification and management. It helps the development team see beyond the many important requirements to isolate and track the few critical requirements that matter the most to successful commercialization.

Conduct Competitive Product Benchmarking

Following the voice of the customer blindly can lead to a failure in the commercialization of a new product. Let's use an analogy to illustrate the point. When a musical band solicits requests from its audience, they had better be able to play the tune! Taking requests needs to be a constrained event unless there is simply nothing you cannot do. One way to establish market segmentation or VOC constraint is by establishing limits on whom you choose to develop and supply products. The limitation is usually set by your company's capability in a given area of technology, manufacturing, outbound marketing, distribution channels, and service capability.

Within your established boundaries of limitation, on a more positive note (some call it a *participation strategy*), you need to be well informed on what your competition is developing and delivering. This requires that some form of competitive benchmarking be conducted. Building a balanced approach to what customers say they want, what you are able to provide, and what your competition is capable of providing is a very reasonable way to conduct product development.

From a Six Sigma performance perspective, we need to decide what critical functional responses need to be at Six Sigma quality. We must use competitive performance data to help define critical functional requirements and then develop designs that can be measured with critical functional responses. This chain of critical requirements and their fulfilling critical responses is the essence of critical parameter management and can be very helpful in defining competitive options. Competitive benchmarking adds one more dimension to what candidate critical requirements and responses need to be considered as we set up the system House of Quality. It can also be one more form of what we call the voice of physics or the voice of the process. If a competitor is doing something new, unique, or difficult relative to what you are currently able to do, you must seriously consider what you will do and what you will *not* do.

The House of Quality has a specific section where you can rank and document these trade-offs. This area of competition vs. your performance trade-off is often where the initial identification of Six Sigma requirements are born. The competition is found to be good at a measurable function and customers say they want more performance in this area, so we have an opportunity to surpass the competition by taking that specific candidate critical function to a whole new level—perhaps to Six Sigma functional performance or beyond. It is important to note that not everything needs to be at a Six Sigma level of performance. If you do find a strategic function that is worth (in terms of value to the business case and the customer) taking to the level of Six Sigma, you'd better be able to measure it against the requirements and against your competition.

Many of the tools of Six Sigma are useful for comparative benchmarking. The following are examples:

- Design and process failure modes and effects analysis
- Taguchi noise diagramming and system noise mapping
- Pugh's concept-evaluation process
- Functional flow diagramming and critical parameter mapping
- Taguchi stress testing
- Highly accelerated life tests and highly accelerated stress testing
- Multivari studies
- Statistical process control and capability studies (for both design functions and process outputs)

Generate Product- or System-Level Requirements Document

The system House of Quality contains the information that relates and ranks the new, unique, and difficult (NUD) VOC needs to the system-level technical requirements that the design-engineering teams must fulfill. These NUD items, along with the many other system-level requirements, must be organized and clearly worded in the system requirements document (SRD). The SRD contains *all* the product-level design requirements. It is important to recognize that three distinct types of requirements must be developed and documented in the SRD:

- Candidate critical system-level requirements from the system House of Quality

- Candidate critical system-level requirements that were not in the system House of Quality because they were not new, unique, or difficult

- Noncritical but nonetheless important system requirements that must be included to adequately complete the product-level design

The system requirements document is the design guide and "bible" that the system-engineering team, subsystem design teams, and all other cross-functional participants in product design and development must seek to fulfill. If a product requirement is not covered in the SRD, it is highly likely that the omitted requirement will not be fulfilled. On the other hand, the SRD is the governing document for what does get delivered. When it is well managed, it prevents design feature and function creep. SRDs are frequently reused and added to as a family of products is developed while the serial flow of product line deployment is conducted.

Design Functions That Fulfill the System Requirements

When the system requirements are clearly worded and documented, the engineering teams can integrate their talents to identify what functions must be developed at the system level to fulfill the system requirements. Defining these functions is the job of a cross-functional team of subsystem engineering team leaders, service engineers,

industrial designers, system engineers, and other appropriate product-development resources, all led by the system engineering and
integration team leader. For organizations that do not have a system-
engineering team leader, the chief engineer or project manager often
leads this effort.

System-level functions are defined as physical performance variables that can be measured to assess the fulfillment of the system
requirements. The functions are ideally defined independent of how
they are accomplished. The focus is on a measurable response that
can be specified as either an engineering scalar or a vector. It is very
important to define continuous variables that will be directly and fundamentally affected by the soon-to-be-defined subsystem functions.
The system functions will have a very strong influence on the definition of the subsystem functions. The subsystem functions will
strongly depend on subassembly and component functions, which, in
turn, are strongly affected by manufacturing process functions.

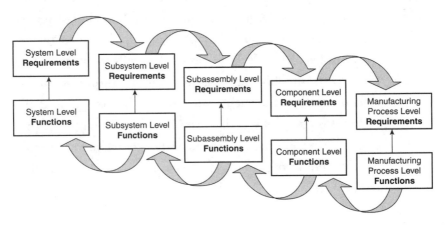

FIGURE 6.6   Critical parameter management model.

Critical parameter management derives from a carefully defined
architectural flow down of requirements that can be directly linked to
functions that are engineered to flow up to fulfill the requirements.

Later we discuss a method of analytically modeling these functional hand-offs across the architectural boundaries within the system. We call these analytical models ideal/transfer functions.

The system-level functions will be used to help define the many possible system architectures that can be configured to fulfill the functions. Here we see how customer needs drive system-level technical requirements. The system-level technical requirements drive the system-level engineering functions. The final link is how the system-level engineering functions drive the definition of the system-level architectural concepts. When a system architecture is estimated from this perspective, the inevitable trade-offs due to subsystem, subassembly, and component architectures begin.

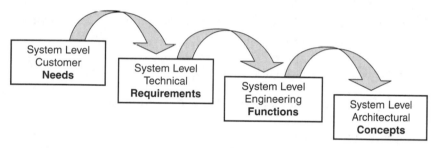

FIGURE 6.7   Critical flow down from needs to concepts.

As you can see, the system drives the design of everything within its architecture.

Generate System Concept-Evaluation Criteria

The criteria used to assess all the candidate system concepts against a "best in class" benchmark must come from the system requirements and system functions. These criteria must be detailed in nature. Stating low cost as a criterion is too general; you must decompose the elements that drive cost and state them as a criterion. This rule applies to most requirements. It is okay to state a system reliability goal, but it might be better to state the more detailed criteria that lead to overall reliability. In this context, low part counts would be reasonable criteria.

The goal of concept development is to define a superior architecture that is high in feasibility and low in competitive vulnerability. A major mistake that many engineering teams make is confusing feasibility with optimization. A major focus in the Concept phase of the product commercialization process is developing the attribute known as *feasibility*. Feasibility is characterized by the following attributes:

- Projected or predicted ability to fulfill a requirement

- Predicted capacity to perform (as opposed to *measured* capability, or Cp, to perform)

- Inherent robustness to customer use variation and environments

- Logical in an application context

- Likelihood or possibility of *balancing* requirements

Engineers are typically undertrained in the tools and best practices for developing feasibility. They tend to have extensive analytical and mathematical training in developing optimality. Feasibility is measured in more generic and general terms against a broad set of system-level requirements. Optimality is typically measured in very specific engineering units against very specific performance targets, usually at the system and subsystem levels. We often find engineering teams overly focused on subsystem optimization much too early in the commercialization process. This leads to subsystems that work well as stand-alone units under nominal laboratory conditions. When these optimized subsystems are integrated into the system configuration, they tend to suffer significant functional performance problems and have to undergo significant redesign efforts. Once in a while, they actually have to be scrapped altogether and a new subsystem concept has to be developed. In terms of cost of poor quality, design scrap and rework costs often have their root cause back in this concept-development phase of commercialization.

Feasibility metrics are focused on requirements that are embryonic and developmental in nature. Requirements will undergo some evolution and refinement as the product commercialization process moves forward. Optimality metrics become clear later in the CDOV process—specifically, in the Optimization phase (Phase 3). Many engineers become impatient and jump the gun on feasibility development and prematurely define just one vague system concept and immediately begin analytical optimization across the various subsystems. The claim is, "We are fast; we get right at the job of engineering the product." The results of taking shortcuts like this are underdeveloped system architectures that are underdefined with regard to their balanced capacity to fulfill all the system requirements and functions. Thus, you can see how a number of subsystems can end up being poorly developed simply due to weakly defined system concept criteria.

Generate System Concepts That Fulfill the Functions

The body of knowledge known as system architecting is comprised of both science and art. System concepts are architected or structured through a set of steps that synthesize the following items:

- System requirements that fulfill the VOC needs

- System functions that fulfill the system requirements

- Engineering knowledge of system design, interface development, and integration

- Previous experience with system performance issues, successes, and failures

The superb reference for conducting this unique discipline, *Systems Architecting*, by Eberhart Rechtin (Prentice Hall, 1990), contains many heuristics or rules of thumb that are extremely useful in guiding the process of defining system architectural concepts.

One of the most import aspects of developing candidate System Concepts is to develop more than one. If just one system concept is developed, it most likely will be lacking in its capability to provide balanced fulfillment of the system requirements. If numerous system concepts are developed and compared to a best-in-class benchmark datum concept, using a comprehensive set of criteria based on the system requirements and system functions, it is much more likely that a superior conceptual architecture can be identified. The power of this approach, known as the Pugh Concept Selection and Evaluation Process, lies in its capability to integrate the best elements from numerous concepts into a supersynthesized superior concept.

This approach drives strong consensus around a team-developed system architecture. It is fine to have a chief system architect, but no one personality should dominate the definition of the actual system concept. One person should lead a disciplined process of system architectural development.

The results of the system-architecting process must come from the team, but the focus on the disciplined use of system conceptual development tools frequently comes from a strong leader of the team. The leader should not dictate a concept, but should insist on using the right tools and process so the team can develop the right concept.

Evaluate System Concepts

System concepts, in the plural, are essential to the proper convergence on a single, superior system concept that will serve as the basis for all subsystem, subassembly, component, and manufacturing process requirements. Too many products have their system defined and driven by the subsystems. This obviously works but is a recipe for system-integration nightmares. So we need to evaluate numerous system concepts.

We must evaluate a number of system concepts developed by individuals or small groups. If one person generates all the concepts, a personal bias or other form of limitation can greatly inhibit the potential to fulfill and balance all the system requirements. You need many concepts with many approaches from diverse perspectives.

Each system concept is compared to the best-in-class benchmark datum using the criteria. Typically between 20 and 30 criteria are used, depending on the nature and complexity of the system requirements and functions. Each team member should bring one or two candidate concepts to the evaluation process. The concepts should be evaluated by using non-numeric standards of comparison:

+ assigned to concepts that are *better* than the datum for a given criterion

− assigned to concepts that are *worse* than the datum for a given criterion

S assigned to concepts that are the *same* as the datum for a given criterion

These evaluation standards efficiently assign value in terms of feasibility. If numbers are used, the evaluation can easily drift away from feasibility and become a typical engineering argument over optimal concepts.

Select Superior System Concept

The evaluation is meant to be iterative. Divergent concepts are now integrated. It is good to step back after a first pass through the evaluation process and integrate the best features and attributes of the various "winning" concepts. This generates a smaller, more potent set of synthesized concepts that can then go through a second round of evaluation. If one of the concepts clearly beats the datum in the first round, it becomes the new datum. The synthesized concepts are then compared to the new datum. After a second round of comparisons is done, a "superior" concept tends to emerge.

A superior concept is defined as the one that scores highest in feasibility against the criteria, compared to the datum. The team should be able to clearly articulate exactly why the selected concept is truly superior to the others and why it is low in competitive vulnerability.

Analyze, Characterize, Model, and Predict Nominal Performance of the Superior System

With a superior concept in hand, the detailed engineering analysis at the system level can begin. It is now reasonable to invest time and resources in system modeling in depth. This is the time when a system-level black-box network model is formally laid out. Here the system-level critical parameter-management model is initially developed.

Up to this point, the critical parameter management database has the structured VOC data (from KJ analysis) residing at the highest level with direct links established to the appropriate system requirements that were developed partly by using the system-level House of Quality from quality function deployment.

The CPM database can now be structured to include the system-level network model of ideal/transfer functions that help represent the sensitivity relationships between system-level requirements and system-level critical functional responses. These are typically estimated first-order equations that enable system-level performance trade-off studies to be quantitatively modeled. These equations express correlations and the magnitude and directionality of sensitivity between system-level CFRs. They are commonly referred to as transfer functions in the jargon of Six Sigma users because they quantify the hand-off or transfer of functional sensitivity. It is important to invest in developing predictions of potentially interactive relationships between the system-level critical functional responses and their requirements. Remember that the "roof" of the system House of Quality should provide our first clue that certain requirements might indeed possess interactive or co-dependent relationships. The system-level transfer functions should account for the actual magnitude and directionality of these predicted co-dependencies. A little later in the Design phase, the subsystem-level sensitivity relationships will be identified as they relate up to the system-level critical functional responses and the system-level requirements they attempt to fulfill.

These system-level transfer functions eventually must be confirmed by empirical methods. This usually occurs during the Design, Optimization, and Capability phases as part of the system engineering and integration process. Designed experiments, analysis of variance, regression, robust design, empirical tolerancing, and response surface methods are the leading approaches for confirming these sensitivity models among the various system-level CFRs.

Develop Reliability Requirements and Initial Reliability Model for the System

Now that we have a system concept defined, a first cut at an integrated model of system requirements and system functions in hand, we can initiate the development of a system reliability model. This model will contain the required system reliability targets and the predicted reliability output at the system level. With system reliability targets and projected reliability performance, the team will have an estimate of the reliability gap between the required and predicted reliability of the system.

This process inevitably requires the superior subsystem concepts to be developed somewhat concurrently with the system so that their effect on the system performance can be estimated. In this context, a subsystem definition has to be considered during system definition.

The hard part is to refrain from letting predetermined subsystem concepts define the system requirements. The system must drive the subsystems—which, in turn, must drive subassembly, component, and manufacturing process requirements. Refinements, trade-offs, and compromises definitely must be made in a concurrent engineering context to keep everything in balance. It is never easy to say exactly what the system requirements are until the subsystems are defined because there will be inevitable dependencies. The system requirements and functions have to change as the subsystem architectures are firmed up. It is crucial to system-architecting and engineering goals to have a strong focus on the system requirements as a *give-and-take* process takes place between system and subsystem architecture and performance trade-offs. The key thing is that system requirements should take precedence over and drive subsystem requirements. This rule can obviously be broken successfully, but in a DFSS context, it should be done only under the constraints and insights from the critical parameter-management process.

Gate 1 Readiness

At Gate 1, we stop and assess the full set of concept-development deliverables and the summary data used to arrive at them. The tools and best practices of Phase 1 have been used, and their key results can be summarized in a Gate 1 scorecard. A checklist of Phase 1 tools and their deliverables can be reviewed for completeness of results and corrective action in areas of risk.

Remember, if you find yourself asking the team why it did not use a specific tool/task grouping, you are out of synch with the order of risk prevention. You should know exactly what tools and tasks were completed. You are merely checking the risk that has accrued in spite of the team's best efforts to prevent problems. Gate reviews deal with issues that happen after well-planned work has been completed properly.

Prerequisite Information to Conduct Phase 1 Activities

- Market segmentation analysis
- Market competitive analysis
- Market opportunity analysis
- Market gap analysis
- Product family plan

General Phase 1 Gate Review Topics

- Business/financial case development
- Technical/design development
- Manufacturing/assembly/materials management/supply chain development
- Regulatory, health, safety, environmental, and legal development
- Shipping/service/sales/support development

General List of Phase 1 Tools and Best Practices

- Market segmentation analysis
- Economic and market trend forecasting
- Business case development methods
- Competitive benchmarking
- VOC-gathering methods
- KJ analysis
- QFD/House of Quality—Level 1 (System level)

- Concept-generation techniques (TRIZ , brainstorming, brain writing, and so on

- Modular and Platform design

- System-architecting methods

- Knowledge-based engineering methods

- Pugh concept evaluation and selection process

- Functional modeling

- Math modeling (business cases, scientific and engineering analysis)

- Reliability modeling

- Design and market FMEAs

- Monte Carlo simulation

- Project cycle-time FMEA

Table 6.1 integrates the Concept phase requirements, deliverables, tasks, and enabling tools.

TABLE 6.1 Tools, Tasks, Deliverables-to-Requirements Integration Table
The CDOV Process and Critical Parameter Management during the Phases and Gates of Product Commercialization

Requirement	Deliverable(s)	Task(s)	Tool(s)
Product requirements derived from current voice-of-the-customer data	Documented voice-of-the-customer data (structured, ranked, and prioritized; KJ document) Documented new, unique, and difficult requirements (product House of Quality) Documented easy, common, and old requirements (general requirements document)	Gather the specific voice of the customer by market segment for the idea bounded by the project being activated from the portfolio. Structure, rank, and prioritize the voice of the customer using KJ analysis. Create product- or system-level House of Quality. Conduct competitive product benchmarking. Generate product- or system-level requirements document.	Market segmentation analysis Economic and market trend forecasting Business case development methods Competitive benchmarking VOC-gathering methods KJ analysis QFD/House of Quality—Level 1 (System level)
Superior product concept derived from a variety of competitive alternatives	Documented system concept alternatives Documented final, superior system (product) concept	Generate system-level concept-evaluation criteria. Generate system-level concepts that fulfill the functions. Evaluate system-level concepts. Select superior system-level concept.	Concept-generation techniques (TRIZ, brainstorming, brain writing, and so on) Modular and platform design System-architecting methods Pugh concept-evaluation and selection process

continues

TABLE 6.1   Tools, Tasks, Deliverables-to-Requirements Integration Table
The CDOV Process and Critical Parameter Management during the Phases and Gates of Product Commercialization
(continued)

Requirement	Deliverable(s)	Task(s)	Tool(s)
Certified technologies from the R&TD process (IIDOV Verify phase)	Documented technology transfer control plans	See IIDOV Verify phase tasks from Chapter 5.	See IIDOV Verify phase tools from Chapter 5.
Functional and reliability models for product/system	Documented system functional models Documented system reliability models	Define the functions that fulfill the system requirements. Analyze, characterize, model, and predict nominal performance of the superior system. Develop reliability requirements, initial reliability model, and FMEA for the system.	Functional modeling Math modeling (scientific and engineering analysis) Reliability modeling Design FMEA Monte Carlo simulation
Technology and market risk profiles	Documented risk profiles for our technology and markets	Generate risk assessment and summary profiles.	Design FMEA Market FMEA Math modeling (business cases) Monte Carlo simulation
Design phase project plans	Documented project plans for the project moving into the Develop phase	Generate Develop phase project plans and cycle-time models.	Monte Carlo simulation Project cycle-time FMEA

The following diagram illustrates the major elements for CDOV.

Product Commercialization

Design Phase: Design Subsystem-, Subassembly-, and Part-Level Elements Based on System Requirements

The Design gate requirements include the following:

- Sublevel design requirements derived from current voice-of-the-customer data, as well as business constraints and technology and competitive benchmarks

- Superior sublevel concept derived from a variety of competitive alternatives

- Certified sublevel technologies from the R&TD process (IIDOV phases)

- Functional and reliability models for sublevel designs

- Baseline capability and critical adjustment parameters for sublevel designs

- Technology, design, and market risk profiles

- Optimize phase project plans

The Design gate deliverables include these:

- Documented voice-of-the-customer data (structured, ranked, and prioritized; KJ document)

- Documented new, unique, and difficult requirements (sublevel Houses of Quality)

- Documented easy, common, and old requirements (general requirements document)

- Documented sublevel functional models

- Documented sublevel design concept alternatives

- Documented final, superior sublevel concepts

- Documented sublevel technology transfer control plans

- Documented baseline design capability indexes under nominal conditions, including critical adjustment parameters (CAPs)

- Documented reliability allocation models for all sublevel designs; sublevel reliability development plans and Design phase results

- Documented risk profiles for new technology, leveraged designs, and markets

- Documented project plans for the projects moving into the Optimize phase

The Design tasks include these:

1. Create sublevel Houses of Quality.

2. Conduct sublevel benchmarking.

3. Generate sublevel requirements documents.

4. Develop functions that fulfill the sublevel requirements.

5. Generate sublevel concept-evaluation criteria.

6. Generate sublevel concepts that fulfill the functions.

7. Evaluate sublevel concepts.

8. Select superior sublevel concepts.

9. Analyze, characterize, model, predict, and measure nominal performance of superior sublevel design concepts (including DFMA, initial tolerances, and cost analysis).

10. Develop a reliability model and FMEA for each sublevel design.

11. Develop Optimize phase project plans.

Transitioning into the second phase of the CDOV product commercialization process requires that the major deliverables from the tools and best practices from Phase 1 are adequately met. It is also predicated on proper advanced planning for use of the Design tools and best practices. This is all part of the preventative strategy of Design for Six Sigma.

Create Sublevel Houses of Quality

It is becoming relatively common to see companies that are deploying DFSS develop a system-level House of Quality. Unfortunately, the use of the lower-level Houses of Quality tends to drop off precipitously after the completion of the system House of Quality. The system House of Quality can and should be used to develop subsystem Houses of Quality for each subsystem within the product architecture. The subsystem House of Quality is extremely useful for developing clear, well-defined subsystem requirements documents. The subsystem requirements must align with and support the fulfillment of the system requirements. In the same context, the subsystem functions must align with and fulfill the system functions. It is our opinion that one of the weakest links in the chain of product commercialization best practices is the cavalier approach design teams take relative to developing clear subsystem, subassembly, component, and manufacturing process requirements. We know from years of experience that it is quite rare to find design teams with clear, well-documented requirements much below the system level. Things tend to get pretty undisciplined and sloppy down in the design details. This is yet another reason why most companies suffer from moderate to severe performance problems that require costly redesign at the subsystem level. The Design phase of CDOV is specifically arranged to instill much-needed discipline in this early, but crucial, area of requirements definition. It is yet another key element in critical parameter management.

At a general level, the subsystem Houses of Quality are designed to illustrate and document specific system requirements as inputs, specifically fulfilling subsystem requirements as outputs and a variety of system-to-subsystem criticality quantification and correlation relationships. They also have "rooms" that quantify competitive benchmarking relationships (from both the customer and engineering team perspectives) as they align with the system and subsystem requirements.

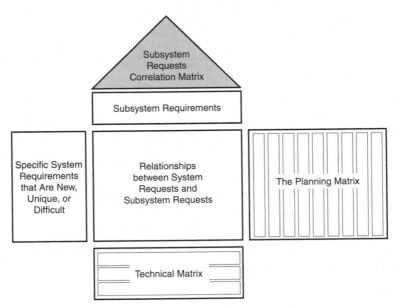

FIGURE 6.8   Subsystem House of Quality.

It bears repeating that only what is new, unique, or difficult from the system requirements should be brought down into the subsystem Houses of Quality.

If the subsystem Houses of Quality are not generated, some form of system integration problem stemming from this lack of assessment and documentation likely will have to be reacted to later in the commercialization process. It just makes good sense to apply the discipline of QFD to the system-to-subsystem requirements that are new, unique, and difficult. Without this focus, these things can transform into that which is expensive, embarrassing, and treacherous. We view the construction of the subsystem Houses of Quality and the subsystem requirements documents as key steps in the DFSS strategy of design problem prevention.

With subsystem requirements defined, the team can proceed into a concurrent engineering mode of design. After structuring the flow-down of the Houses of Quality, the development of subassemblies, components, and manufacturing processes can proceed in any mix of serial and parallel activities, based on tools and best practices that make sense to the subsystem design teams. The design teams must resist putting the cart before the horse. Short-circuiting the development of the subordinate Houses of Quality adds risk to your team. You are highly likely to miss something relative to what is new, unique, or difficult down in the details of the subsystem designs. The overarching rule for conceptual development is to let the hierarchical flow-down of system-to-subsystem-to-subassembly-to-component-to-manufacturing process requirements lead to the sequencing of concept generation, evaluation, and selection. The design development tasks then become a series of trade-offs made on a daily basis between the form, fit, and functions within and between the hierarchy of design elements. The critical parameter-management process, under the leadership of your system-engineering team and its process, is your guide to balancing sensitivities and solving the many-to-many parametric problems you will encounter within and between the subsystems and their elements.

Conduct Sublevel Benchmarking

When appropriate, subsystems should be benchmarked from the customer's perspective. If the product design is such that the customer has no possible access to or interest in one or more of the subsystems, the benchmarking activity will be solely conducted by the engineering team from a technical perspective. If the product is serviceable or repairable, the benchmarking perspective of the "customer in the form of the repair technician" is required.

A clear case in which subsystem benchmarking from a customer's perspective would be necessary arises in the replacement process of inkjet cartridges in a personal desktop printer. On the other hand, a customer would have little role in subsystem benchmarking regarding the efficiency of a sealed cooling subsystem within a light projector used for business presentations.

Subsystem benchmarking is important because it adds information that can be used to help establish which subsystems possess more critical, value-adding functions, compared to those that are of less consequence. The customer benchmarking input helps the team identify advantages in competitive features. The technical benchmark data can uncover performance or architectural design information that can be targeted, surpassed, or exploited if a weakness is found. Benchmarking essentially is all about exposing vulnerabilities and opportunities. The ensuing concept-generation, evaluation, and selection processes are all about averting, minimizing, or at least recognizing competitive vulnerabilities. Sometimes it is wise to allow yourself to be equaled or surpassed by a competitor in certain areas that are not strategic to your business case. In the copier industry, Six Sigma quality is strategic in the area of image quality but not in cabinetry fit and finish (at least, not yet). The auto industry is another matter: Fit and finish of interior and exterior body elements is a matter for Six Sigma performance metrics.

Now that we are getting down into the details of the subsystems and their subordinate requirements, architectures, and functions, not every subassembly, component, or manufacturing process needs to go through competitive or comparative benchmarking. This becomes more of a judgment call within the context of what is deemed critical to function and reliability for a given cost.

As part of the critical parameter management work, the Houses of Quality and the resulting requirements help make it apparent when a subassembly, component, or manufacturing process is critical enough to require the benchmarking of alternatives for your consideration. When the subassembly or component can be effectively outsourced, benchmarking should certainly be considered. Outsourcing should be done when a supplier can make designs or components better and cheaper than you can. Many companies are reducing their vertical integration, some for good reasons—some are walking away from things that should remain core competencies. Making certain critical subassemblies and components in-house is a strategic decision, which DFSS and critical parameter management can and should greatly influence. Developing specific capability in certain critical manufacturing and assembly processes falls into the same scenario. Benchmarking, in the context of Design for Six Sigma and Production Operations for Six Sigma, will help focus what you choose to keep inside your company and what to safely outsource. Great care should be exercised in the sourcing of critical-to-function or critical-to-reliability subassemblies and components.

Benchmarking is often used to help identify good ideas for how competitors develop functions, as well as how they develop form and fit architectures that generate functions. Benchmarking architectures is common, but we also highly recommend separating the form from the function so your team gets a good sense of both. We have seen successful (revenue-bearing) patents come from copying competitive functions without copying their form and fit.

Generate Sublevel Requirements Documents

We cannot overemphasize the importance of creating clear, well-defined subsystem requirements documents. This is done by combining the subsystem House of Quality data with all the other requirements that are associated with each of the subsystems that comprise the integrated system. The subsystem Houses of Quality should not contain all the subsystem requirements; they should contain just those requirements that are new, unique, or difficult. The subsystem teams are responsible for developing designs that will fulfill all of the requirements—those that are critical to the newness, uniqueness, and difficulty requirements that have the largest impact on fulfilling the VOC needs, as well as the rest of the requirements that are of general importance. If a subsystem is being reused from a previous product, many of the requirements will simply be repeated from the last design. A small set of additional or upgraded requirements will be added to the list of reused requirements. All requirements are important, but a few are absolutely critical in identifying the subsystem's functional performance. The team might not know exactly which ones are "truly critical," but it's a good bet that many will come from the subsystem Houses of Quality.

It is just as important to develop and fulfill a comprehensive set of requirements at this detailed level of design and manufacturing as it is at the system and subsystem levels. The network of critical parameter relationships can get quite complex down at this level. In our experience, design teams relax the rigor of requirement definition after the system-level requirements document is produced. Some of the biggest problems that result from this lack of follow-through on requirements definition are anemic metrics and measurement systems. When subassemblies, components, and manufacturing processes are undermeasured, they tend to underperform later in the system integration, production certification, and customer utilization phases of the product's lifecycle. So the system requirements often are difficult to fulfill because the design elements that must support them are underdefined and, consequently, incapable of delivering their portion of required performance. Each of these three design elements should have requirements documents defined from a blend of the new, unique, or difficult requirements that translate out of their respective Houses of Quality, along with all other important requirements.

Develop Functions That Fulfill the Sublevel Requirements

The subsystem teams need to assign a means of measuring the functions that prove that a requirement has been fulfilled. These functions are identified independent of the concepts that will soon be generated to fulfill them. When a subsystem concept is defined, the final critical functional responses can be identified along with their specific data-acquisition methodology.

Defining a generic set of subsystem functions initiates another key element of the critical parameter management process. The generic set of functions implies a system of metrics within each subsystem that will eventually evolve into the hardware, firmware, and software used to actually measure the nominal performance, robustness, tunability, and capability of the subsystem.

The flow of functions that are essential to the fulfillment of the hierarchy of the product's requirements must be supported by these three elements of the design. Manufacturing processes provide functions to convert mass and energy into components. The components, in turn, provide an integrated set of contributions in support of subassembly and subsystem functionality.

Subassembly functions are developed based on the subassembly requirements. They, too, must indicate how the requirements will be fulfilled without specifically stating how the function will be accomplished. Subassemblies are typically required to create a mechanical, electrical, or chemical function based on an integrated set of components.

Components are a little different because they do not, by themselves, transform or change the state of energy and mass. They must be integrated into a subassembly or subsystem to add up to the accomplishment of the subassembly or subsystem functions. When we discuss "additivity" in the robust design material, it directly relates to this notion of design elements "adding up" their contribution to facilitate functional performance. Components tend to possess parametric characteristics that are best quantified under some form of specification nomenclature. This is why we call critical functional forms, fits, and features of a component critical-to-function specifications. These typically are documented as dimensions or surface or bulk characteristics. A structural component might have a deflection requirement that is fulfilled through its capability to provide stiffness. This does not suggest how the component will provide stiffness—just that whatever component architecture is selected must be capable of being measured in the units of stiffness to fulfill a deflection requirement (area moment of inertia [in^4] controlling a deflection [in.]).

Manufacturing processes create functions to make components. They control mass and energy states, flows, and transformations that ultimately become components. If the functions within a manufacturing process are directly linked to the manufacturing process requirements, which came from component, subassembly, and subsystem requirements, we have a much better chance of flowing functional integrity up through the system. How often do you link and relate the functions inside your production process to the functions within your designs?

Generate Sublevel Concept-Evaluation Criteria

Concept-evaluation criteria, which is used in the Pugh concept-selection process, must be based on the subsystem requirements. Typically, the concept criteria is detailed in nature. The team frequently breaks general or broadly-stated requirements into two or more focused or descriptive criteria statements that provide enough detail to discriminate superiority and low vulnerability between the datum concept and the numerous candidate subsystem concepts. As was the case in system criteria generation, feasibility—not optimality—is being evaluated. The scope and depth of the criteria need to reflect this goal. It is okay to develop the evaluation criteria before, during, and after the concepts have been generated. In fact, quite commonly, criteria are added, removed, or modified during the concept-evaluation process.

Criteria to evaluate the subassembly, component, and manufacturing process concepts must be developed at this point. It is not uncommon to develop the criteria during concept generation. The subassembly, component, and manufacturing process requirements documents are used to help define the concept-evaluation criteria.

Generate Sublevel Concepts That Fulfill the Functions

This step defines the "form and fit" options that can potentially fulfill the requirements and functions. These are commonly known as subsystem concepts; they are also referred to as candidate subsystem architectures. As is always the case in concept development, the team needs to generate numerous subsystem concepts to be compared to a "best-in-class" benchmark datum.

Concept-generation tools range from group-based brainstorming techniques to individual methods that help stimulate creativity in developing concepts. TRIZ is a computer-aided approach to stimulating inventions and innovations. A lone innovator often uses TRIZ when developing numerous concepts. Individuals commonly do concept generation, whereas concept evaluation is done in small groups (of seven to ten people). It is important to set some standards for describing and illustrating the concepts. It is undesirable to have some concepts that are underdefined or poorly illustrated while others are highly defined and clearly illustrated. Therefore, the team must agree on a common standard of thoroughness and completeness of concept documentation. Areas to consider are as follows:

- Textual description of the concepts
- Graphical representation of the concepts

- Mathematical equations that underwrite the basic principles of functional performance

- Physical models of the concepts

However the team chooses to document each concept, everyone must adhere to reasonably similar representations, for the sake of fair comparisons.

With the detailed functions outlined, we can effectively generate the concepts for the subassemblies, components, and manufacturing processes. This process tends to go faster than system and subsystem concept generation because there are fewer, less complex requirements down at this level. The tools and best practices to develop these concepts are linked to the physical and structural nature implied in the requirements and functions. Creativity, brainstorming, knowledge-based engineering-design methods, and TRIZ are a few examples of the tools used to generate concepts at this level. The need for numerous concepts is just as important as it was with the higher-level subsystem and system concepts.

Evaluate Sublevel Concepts

Each subsystem concept is compared to the "best-in-class" benchmark datum using the criteria. Typically between 20 and 30 criteria are used, depending on the nature and complexity of the subsystem requirements and functions. Each team member should bring one or two concepts to the evaluation process. The concepts should be evaluated by using non-numeric standards of comparison:

+ assigned to concepts that are *better* than the datum for a given criterion

− assigned to concepts that are *worse* than the datum for a given criterion

S assigned to concepts that are the *same* as the datum for a given criterion

These evaluation standards are efficient for the purpose of assigning value in terms of feasibility. If numbers are used, the evaluation can easily drift away from feasibility and become a typical engineering argument over optimal concepts.

The Pugh Concept Evaluation and Selection Process is a common and popular tool used to arrive at a set of superior concepts. This approach helps prevent designs and production processes from being developed or selected with incomplete forethought. Many times it is tempting to just go with one person's gut feel or experience. This has proven over time to be an incomplete approach to concept selection. When a cross-functional team brings its accumulated knowledge, experience, and judgment to bear on a number of concepts, it is much more likely that a superior concept will emerge from the concept-evaluation process. Integrating the best elements of numerous concepts into a design that is low in competitive vulnerability and high in its capability to fulfill the criteria is by far the best way to help drive the design to fulfill its requirements.

Select Superior Sublevel Concepts

The numerous divergent subsystem concepts are compared to the datum during the first round of evaluation. The subsystem concepts that score high in the comparison can have their best attributes integrated with similar concepts. This concept-synthesis process causes a convergence of criteria-fulfilling features, functions, and attributes that often enable the team to find one or more concepts that surpass the datum. One of the superior candidate concepts is used to replace the original datum. A second round of evaluation is then conducted, comparing a smaller, more powerful group of supersynthesized subsystem concepts to the new datum. When a clearly superior subsystem concept emerges, it is now worth taking forward in the design process. If the concept-evaluation process is bypassed, it is highly likely that the subsystem that is taken into design will suffer from an inability to provide complete and balanced fulfillment of the subsystem requirements. It is easy to go with one pet concept to the exclusion of generating any others, but it is very hard to fix such a concept's conceptual vulnerabilities later.

The final selection of the superior subassembly, component, and manufacturing process concepts often requires many compromises as design functionality vs. produceability trade-offs are made in a concurrent engineering context. As long as the criteria can sufficiently account for the range of requirements for these three elements, the likelihood of the most feasible designs and processes being selected is high.

Just as it is important to evaluate numerous concepts, it is important to be decisive and go forward with confidence in the selected design or process. When two relatively equal concepts come out of the selection process, it might be wise to co-develop both concepts until a clear winner emerges during the more formal development of the models of the concepts. Data is the tiebreaker.

Analyze, Characterize, Model, Predict, and Measure Nominal Performance of Superior Sublevel Designs (including DFMA, Initial Tolerances, and Cost Analysis)

With a superior set of subsystem concepts defined, it is now time to add a great deal of value to them. They are worth the investment that will be made in them through the application of a number of tools and best practices. In traditional Six Sigma methods, the Measure, Analyze, Improve, and Control process is applied to previously designed and stabilized production processes. In new or redesigned subsystems, the approach is different. The subsystem designs are not yet statistically stable "processes" in the context of Walter Shewhart's approach to statistical process control. In fact, they are just subsystems within a system design, not production processes. The DMAIC approach has its historical basis in manufacturing process control charting based on the fact that a production process is in a state of "statistical control." This assumes that all assignable causes of variation are known and removed. It also assumes that the variation in the function of the process is due to random, natural causes that are small in comparison to the "assignable, non-natural causes (the kind induced in a Taguchi Robustness Experiment)" of variation.

We simply don't know enough about the subsystem parameters and measured responses to state whether the subsystem is under statistical control from a steady-state production point of view. We have design and optimization work to do on the new subsystem concepts before they are ready for the type of "long-term" capability characterization that comes from the traditional DMAIC process. During the Design phase, it is possible to generate and track short-term design capability performance. During the Optimize and Capability phases, it is possible to emulate long-term shifts by intentionally inducing special cause sources of variation using Taguchi methods. In this sense, the design team can simulate Cpk data. One must recognize that this is a very good thing to do—but it is only a simulation of the actual variation to come when real production and customer use begin.

The DMAIC process fits very nicely within the Verify phase of the CDOV commercialization process. By then the subsystem designs are mature enough to be characterized using the DMAIC steps

- Defined set of corrective actions that must be closed out before launch

- Measured against final design requirements

- Analyzed for statistical performance in a production context

- Improved by adjusting the mean of the design critical functional responses onto the desired target using critical adjustment parameters that were developed during the Design and Optimization phases of CDOV

- Controlled under the designed plan based upon critical parameters during the steady-state production, service, and customer use processes

Analysis, Characterization, and Modeling in the Design Phase

Each subsystem is ready to be formally analyzed, characterized, and modeled. This is done using a blend of knowledge-based engineering tools and best practices, as well as the Design for Six Sigma tools and best practices. The chief goals within the Design phase are to analytically predict and then empirically measure the nominal or baseline performance of the superior subsystems. Here the subsystem critical parameters are identified and developed. The modeling focuses on developing basic relationships, called ideal/transfer functions. These are linear and nonlinear systems of equations that are comprised of dependent design parameters, (Y) variables and independent variables, and the subsystem (x) variables. Actually, the models also must account for the fact that some of the (x) variables might not be totally independent from one another in their effect on the (Y) variables. We call these co-dependent "x-to-x" variable relationships *interactions.* The most popular method of developing these ideal/transfer functions is to use a blend of analytical modeling methods and experimental modeling methods.

Measurement Systems in the Design Phase

Probably the most value-adding feature that Six Sigma methods bring to the Design, Optimization, and Capability phases of product commercialization is its rigorous approach to measurement. Within DFSS, the critical parameter management approach leads one to focus on the measurement of critical functional responses (usually in engineering units of vectors—magnitude and direction) and critical-to-function specifications (usually in engineering units of scalars—magnitude only).

Before the evolving subsystems can be experimentally evaluated, they must undergo detailed development for data acquisition. Design for Six Sigma builds value and prevents problems by measuring things that avoid the cost of poor quality. The root causes of the cost of poor quality reside in the capability of the mean and standard deviations of the critical functions to possess low variability while maintaining on-target performance. You will have a hard time attaining that kind of performance by measuring defects. What is the meaning of the mean and standard deviation of a defect? The design teams must invest in the development and certification of measurement systems that provide direct measures of functional performance. We cannot afford to wait until a failure occurs to assess design quality. We must instill discipline in our teams to avoid measures of attributes, yields, go-no events, defects, and any other measure of "quality" that is not fundamental to the flow, transformation, or state of energy and mass within the function of the design.

Measurement system development and analysis is a key tool in DFSS. This verifies the capability of instruments that are designed to measure continuous variables that represent the functions of the design, as opposed to the more technically ambiguous measures of quality after the function is complete. We caution against waiting to measure design performance after all the value-adding functions are over. When we link this function-based metrology approach to designed experimentation, very powerful results based on functional data are delivered at the CDOV gate reviews.

Statistically designed experiments (called *design of experiments,* DOE) are some of the most heavily used tools from the methods of Six Sigma. An approach known as sequential experimentation can efficiently and effectively develop critical parameter knowledge and data:

- **Screening experiments** define basic, linear, main effects between a Y variable and any number of x values. This approach usually just looks at the effects of independent x variables.

- **Main effects and interaction identification experiments** define how strong the main effects and the interactions between certain main effects (x's) are on the Y variables.

- **Nonlinear effects experiments** use a variety of experimental data-gathering and analysis structures to identify and quantify the strength of nonlinear effects of certain x variables (second-order effects and beyond as necessary).

- **Response surface experiments** study relatively small numbers of x's for their optimum set points (including the effects from interactions and nonlinearities) for placing a Y or multiple Ys onto a specific target. They can also be used to optimally reduce variability of a critical functional response in the presence of noise factors, but only on a very limited number of x parameters and noise factors.

These four general classes of designed experimentation somewhat understate the power resident in the vast array of designed experimental methods that are available to characterize, model, and enable many forms of analysis for the basic and advanced development of any subsystem, in terms of both mean and variance performance.

Produceability in the Design Phase

Included in the design characterization and analysis work is the process of designing for manufacturability and assembleability, the development of the initial functional and assembly tolerances that affect the output of the subsystem critical functional responses and, of course, design cost analysis.

Even though some consideration for manufacturing and assembly is necessary during the subsystem concept-generation process, there is always more to do as the Design phase adds depth and maturity to the subsystems' form, fit, and function. Formal application of Design for Manufacturability and Assembleability methods are essential elements at this point in the progressive development of each subsystem design. This is also an excellent point in the commercialization process to establish initial tolerances along with a detailed cost estimate for each subsystem.

Design and manufacturing tolerances are embryonic at this point in the CDOV process; they are mainly estimates. Relationships within and between subsystem functions are being quantified and their transfer functions being documented. Relationships with subsystem, subassembly, and component suppliers are also in their early stages of development. Two classes of tolerances need to begin to be balanced to have a chance at achieving affordable Six Sigma performance: functional performance tolerances on critical functional responses within and between subsystems and subassemblies, and assembly/component tolerances. One class focuses on functional sensitivities and the tolerances required to constrain functional variation. The other class focuses on form and fit sensitivities and the tolerances that limit that balance of components and subassemblies as they are assembled into subsystems. Subsystem-to-system integration tolerances also must be projected and analyzed for form, fit, and function at this point. Initial tolerances for both classes can be established using the following analytical tolerance design tools and best practices:

- Worst-case tolerance studies

- Root sum of squares tolerance studies

- Six Sigma tolerance studies

- Monte Carlo simulations

It is also a very opportune time to begin to build a detailed database of supplier manufacturing and assembly process capabilities (Cp and Cpk) to minimize the risk of a process-to-design capability mismatch. Here again, we see the preventative, proactive nature of the Design for Six Sigma process.

The analysis, characterization, modeling, predicting, and measurement of the various concepts is usually developed in the following general order (math and geometric modeling are often done in parallel as they support and enable one another):

- **Math models** (functional performance, tolerance stack-ups, cost estimates)

- **Geometric models** (spatial form and fit and tolerance stack-ups)

- **Physical models** (prototypes to evaluate function and fit)

The most common tools from DFSS in support of this activity are

- Design for manufacture and design for assembly

- Conduct measurement system analysis (design for testability)

- Design experiments, ANOVA data analysis, and regression (deriving $Y = f(x)$ equations)

  - Ideal/transfer functions within and between subsystems and subassemblies

- Institute critical parameter management

  - Functional flow diagramming

- Design capability studies (baseline Cp characterization of CFRs)

- Design applications of statistical process control (nominal performance stability)

- Multivari studies (screening for "unknown" design influence variables)

- Value engineering and analysis (cost-to-benefit analysis and FAST diagramming)

- Analytical tolerance development and Simulation (*ideal/ transfer functions*)

A great deal of trade-off work is done in this step. Refinement of the initial concepts is done continuously as costs, function, and pro-duceability requirements are initially balanced. Usually the designs and processes are set at the most economical level possible at this point. If basic (nominal) functionality, manufacturability, and assem-bleability (the latter two terms we often refer to jointly as *produce-ability*) are attained, we no longer refer to these elements as concepts: They are then full-fledged designs and processes. The Optimization phase of CDOV attempts to develop robustness (insen-sitivity to "assignable cause" sources of variation) in the designs and processes without significant increases in cost. The Capability phase of the CDOV process might require significant increases in cost to attain the proper performance and reliability requirements within the scheduled delivery time of the product to the market.

Develop Reliability Model and FMEA for Each Sublevel Design

The system reliability model, from Phase 1, should contain an overall reliability growth factor based on the projected reliability gap between required system reliability and the current system reliability. It is common practice to derive a reliability allocation model from which a reliability budget flows down to each major subsystem. Each subsystem needs to be analyzed for its projected capability to meet the system budget. It is then possible to define a subsystem reliabil-ity gap that must be closed for the system to meet its goals.

After the subsystem reliability requirements are documented, the team must embark on defining where the reliability-critical areas reside within the subsystems. A strong method for identifying critical reliability areas in any design is called *design failure modes and effects analysis* (DFMEA). A more rigorous form includes a probability of failure value known as *criticality*. Thus, we see that the DFMECA method includes a *probability of occurrence* term. The DFMEA data helps the team predict the reliability of the developing subsystem. It highlights the likely failure modes and the specific effect they will have on the reliability of the subsystem and, ultimately, on the system. DFMEA data is useful not only in defining preventative action in the development of reliability, but also in the process of developing robustness, as discussed in the Optimization phase.

It is worth stating that, in the context of the CDOV process, DFSS has an extremely aggressive approach to developing reliability. The CDOV process contains a heavy focus on reliability requirement definition, modeling, and prediction at the front end of the process and a strong focus on assessing attained reliability at the tail end of the process. The tools and best practices in the Design and Optimize phases also do a great deal to develop and measure surrogate, forward-looking measures of performance that directly affect reliability (means, standard deviations, signal-to-noise metrics, and Cp/Cpk indexes). Waiting to measure mean time to failure and other time-based measures of failure is not an acceptable Six Sigma strategy. DFSS is not fundamentally reactive. We do everything possible at every phase in the CDOV process to get the most out of the DFSS tools and best practices to prevent reliability problems.

Concurrent with the development of the designs and processes, the engineering team should be constructing reliability performance predictions. These predictions are used to roll up estimated reliability performance to fulfill overall system reliability requirements.

Reliability allocations are made down to the subsystem, subassembly, and component levels within the system architecture.

As the prediction process moves forward, shortfalls or gaps always will emerge in predicted vs. required reliability. Design failure modes and effects analysis is a major tool used in the very earliest stages of reliability prediction and development to identify areas of weakness in the designs. It also provides insight into exactly what is likely to cause a reliability problem. If DFMEA is done early in the design process, the team gains a strategic capability to plan and schedule the deployment of numerous reliability development tools and best practices to prevent reliability problems before they get embedded in the final design. A list of DFSS reliability development tools and best practices follows:

1. Conduct DFMEA.

2. Develop a network of ideal and transfer functions using math modeling and DOE methods.

3. Conduct sensitivity analysis on the ideal and transfer functions using Monte Carlo simulations.

4. Conduct Taguchi's robustness-optimization process on sub-assemblies and subsystems prior to system integration (optimized Cp).

5. Use response surface methods to identify critical adjustment parameters to adjust the design functions to their performance targets, to maximize Cpk.

6. Do normal life testing.

7. Do highly accelerated life-testing (HALT).

8. Do highly accelerated stress testing (HAST).

At this point in time, conducting reliability assessments on reliability critical subassemblies and components is appropriate if they have already been through the supplier's robustness-optimization process. If the supplier refuses to put them through robustness optimization, it is highly advisable that you begin life-testing "yesterday." If the supplier offers no proof of confirmed robustness but says it did so, start life-testing immediately. If the supplier agrees to conduct robustness optimization on the subassembly or component, help the supplier do it; then, when you are sure the design has been optimized, you can conduct life-testing procedures as necessary. It is generally a waste of time to conduct life tests on designs that will soon go through robustness optimization—wait until that is done and then go forward with the life-testing.

Design Process Steps Down to the Subassembly, Component, and Manufacturing Process Levels

To emphasize the hierarchical nature of sublevel design iteration, we include the key tasks that we have just developed with an emphasis on the subassembly, component, and manufacturing process levels.

1. Create subassembly, component, and manufacturing process Houses of Quality.

2. Conduct subassembly, component, and manufacturing process benchmarking.

3. Generate subassembly, component, and manufacturing process requirements documents.

4. Develop functions that fulfill the subassembly, component, and manufacturing process requirements.

5. Generate subassembly, component, and manufacturing process-evaluation criteria.

6. Generate subassembly, component, and manufacturing process concepts that fulfill the functions.

7. Evaluate subassembly, component, and manufacturing process concepts.

8. Select superior subassembly, component, and manufacturing process concepts.

9. Analyze, characterize, model, predict, and measure nominal performance of superior subassembly, component, and manufacturing processes (including DFMA, initial tolerances, and cost analysis)

10. Develop reliability models and FMEAs for reliability critical subassemblies and components

If you are wondering what is taking your commercialization teams so long to get a product launched, it is probably because they are rushing. We have found that rushing through the phases of work without careful selection of value-adding tasks enabled by a rational set of tools leads to a lot of design scrap and rework. Time just evaporates in an extra phase of work that lies hidden in the final Verify phase. Here all the mistakes and omissions (a classic Six Sigma person would call these escaping defects) from rushing come to a head. The most debilitating of these get beaten down to an acceptable level (typically to management, not the customer), stretching the schedule way beyond what was promised, and then the team launches a relatively unfinished product. It finally gets completed during the first year of sales. Ironically, many companies use DMAIC projects to clean up the mess. Don't rush—do it right the first time by designing a linked set of tool/task groups to prevent latent, post-launch DMAIC projects.

Gate 2 Readiness

At Gate 2, we stop and assess the full set of deliverables and the summary data used to arrive at them. The tools and best practices of

Phase 2 have been used, and their key results can be summarized in a Gate 2 scorecard. The Phase 2 tool/task groups and their deliverables can be reviewed for completeness of results and corrective action in areas of risk.

Phase 2 Gate Review Topics

- Business/financial case development

- Technical/design development

- Manufacturing/assembly/materials management/supply chain development

- Regulatory, health, safety, environmental, and legal development

- Shipping/service/sales/support development

A General List of Phase 2 Tools, Methods, and Best Practices

- Competitive benchmarking

- VOC-gathering methods

- KJ analysis

- QFD/Houses of Quality—Levels 2 to 5 (subsystem, subassembly, component, and manufacturing process levels)

- Concept-generation techniques (TRIZ, brainstorming, brain writing, and so on)

- Pugh Concept Evaluation and Selection Process

- Design for manufacture and assembly

- Value engineering and analysis

- Design failure modes and effects analysis

- Measurement systems analysis

- Critical parameter management

- Knowledge-based engineering methods

- Math modeling (business case and engineering analysis to define ideal functions and transfer functions)

- Design of experiments (full and fractional factorial designs, sequential experimentation)

- Descriptive and inferential statistical analysis

- ANOVA data analysis

- Regression and empirical modeling methods

- Design capability studies

- Multivari studies

- Statistical process control (for design stability and capability studies)

- Reliability modeling methods

- Normal and accelerated life-testing methods

- Monte Carlo simulation

- Design FMEA

- Project cycle-time FMEA

Table 6.2 integrates the Design phase requirements, deliverables, tasks, and enabling tools.

TABLE 6.2 Tools, Tasks, Deliverables-to-Requirements Integration Table

Requirement	Deliverable(s)	Task(s)	Tool(s)
Sublevel design requirements derived from current voice-of-the-customer data, as well as business constraints and technology and competitive benchmarks	Documented voice-of-the-customer data (structured, ranked, and prioritized; KJ document) Documented new, unique, and difficult requirements sublevel Houses of Quality) Documented easy, common, and old requirements (general requirements document)	Create sublevel Houses of Quality. Conduct sublevel benchmarking. Generate sublevel requirements documents.	Competitive benchmarking VOC-gathering methods KJ analysis QFD/Houses of Quality—Levels 2 to 5 (subsystem, subassembly, component, and manufacturing process levels)
Superior sublevel concept derived from a variety of competitive alternatives	Documented sublevel design concept alternatives Documented final, superior sublevel concepts	Generate sublevel concept-evaluation criteria. Generate sublevel concepts that fulfill the functions. Evaluate sublevel concepts. Select superior sublevel concepts.	Concept-generation techniques (TRIZ, brainstorming, brain writing, and so on) Pugh Concept Evaluation and Selection Process
Certified sublevel technologies from the R&TD process (IIDOV phases)	Documented sublevel technology transfer control plans	See Chapter 5 for IIDOV tasks associated with this work.	See Chapter 5 for IIDOV tasks associated with this work.
Functional and reliability models for sublevel designs	Documented sublevel functional models Documented reliability allocation models for all sublevel designs; sublevel reliability development plans and Design phase results	Develop functions that fulfill the sublevel requirements. Develop reliability model and FMEA for each sublevel design.	Functional modeling Reliability modeling methods Normal and accelerated life-testing methods Monte Carlo simulation Design FMEA

Inputs	Deliverables	Tasks	Tools and Methods
Baseline capability and critical adjustment parameters for sub-level designs	Documented baseline design capability indexes under nominal conditions, including critical adjustment parameters (CAPs)	Analyze, characterize, model, predict, and measure nominal performance of superior sublevel design concepts (including DFMA, Initial Tolerances & Cost Analysis)	Design for manufacture and assembly Value engineering and analysis Design failure modes and effects analysis Measurement systems analysis Critical parameter management Knowledge-based engineering methods Math modeling (business case and engineering analysis to define ideal functions and transfer functions) Design of experiments (full and fractional factorial designs, sequential experimentation) Descriptive and inferential statistical analysis ANOVA data analysis Regression and empirical modeling methods Design capability studies Multivari studies Statistical process control (for design stability and capability studies)
Technology, design, and market risk profiles	Documented risk profiles for new technology, leveraged designs, and markets	Generate risk profiles for new sublevel technologies, leveraged sublevel designs and market.	Design FMEA Market FMEA
Optimize phase project plans	Documented project plans for the projects moving into the Optimize phase	Develop Optimize phase project plans.	Monte Carlo simulation Design FMEA Project cycle time FMEA

The CDOV Process and Critical Parameter Management during the Phases and Gates of Product Commercialization

Optimize Phase: Optimize Sublevel Designs and the Integrated System

Product Commercialization

Phase 3 contains two major parts:

- Phase 3A focuses on the robustness optimization of the individual subsystems and subassemblies within the system.

- Phase 3B focuses on the integration of the robust subsystems and subassemblies, the nominal performance evaluation, robustness optimization, and initial reliability assessment of the complete system.

The Optimize gate requirements include the following:

- Sublevel designs that are robust and tunable

- Sublevel design reliability growth

- Sublevel design risk assessment

- System that is robust and tunable

- System design reliability growth

- System risk profiles

- Critical Parameter-Management Database

- Verify phase project plans

The Optimize gate deliverables include these:

- Documented robustness and tunability of each sublevel design (at functional assembly level and higher)

- Documented reliability growth of all sublevel designs (under nominal and stressful conditions)

- Documented reliability growth of the integrated system (under nominal and stressful conditions)

- Documented robustness and tunability of the integrated system

- Documented risk assessment of the sublevel designs

- Documented risk profile of the system

- Documented sublevel critical parameters

- Documented system-level critical parameters

- Documented project plans for the Verify phase

The Optimize tasks for Phase 3A include these:

1. Review and finalize the critical functional responses for the subsystems and subassemblies.

2. Develop subsystem noise diagrams and system noise map.

3. Conduct noise factor experiments.

4. Define compounded noises for robustness experiments.

5. Define engineering control factors for robustness optimization experiments.

6. Design for additivity and run designed experiments or simulations.

7. Run verification experiments to verify robustness.

8. Analyze data, build predictive additive model.

9. Conduct response surface experiments on critical adjustment parameters.

10. Run verification experiments to verify tunability and robustness parameters for subsystems and subassemblies

11. Document critical functional parameter nominal set points; CAP and CFR relationships.

12. Develop, conduct, and analyze reliability/capability evaluations for each subsystem and subassembly.

Review and Finalize the Critical Functional Responses for the Subsystems and Subassemblies

Before investing a lot of time and resources in conducting the steps of robust design, it is important to step back and review all the critical functional responses for the subsystems and subassemblies. These direct measures of functional performance tell the teams whether their designs are sensitive to sources of variation. Each CFR must have a capable measurement system that can be calibrated and certified for use during the optimization phase of the commercialization process. Again, we emphasize that the CFRs being measured should be continuous engineering variables. They should be scalars (magnitude of the measured variable) and vectors (magnitude and direction of the measured variable) that are capable of expressing the direct flow, transformation, or state of mass, energy, and the controlling signals that reflect the fundamental performance of the designs being optimized.

Robust design is all about how the CFRs react to sources of noise (variation) for a given set of control factor set points. These control factor set points will become part of the final design specifications when optimization is complete. It is very important that their relationships to the noise factors be evaluated and exploited. Noise factors are things that are expensive, difficult, or even impossible to control. A special, often ignored class of interactions between control factors and noise factors shows up in the measured CFRs at the system, subsystem, and subassembly levels in all products. The purpose of robust design is to investigate, quantify, and exploit these unique control factor-to-noise factor interactions. Measuring defects, yields, and go/no-go quality metrics is inappropriate for this physical engineering context. Recall that it is quite rare to hear a lecture in physics, chemistry, or just about any engineering topic centered on quality metrics—they all focus on basic or derived units that we refer to as engineering metrics (scalars and vectors). This is how you must approach DFSS metrics for critical parameter management and robust design, in particular.

Develop Subsystem Noise Diagrams and System Noise Map

Identifying noise factors is a key step in the optimization of any system. A noise factor is anything that can induce variation in the functional performance of a design. The critical functional response is the variable used to measure the effects of noise factors. Noise factors are typically divided into three general categories:

- **External noises**—Variation coming into the design from an external source

- **Unit-to-unit noises**—Variation stemming from part-to-part, batch-to-batch, or person-to-person sources

- **Deterioration noises**—Variation due to wear or some other form of degraded state or condition within the design

A noise diagram can be constructed to illustrate the noise factors that are thought to have a large impact on the CFRs for any given subsystem or subassembly. It is most often used to aid in structuring two types of designed experiments:

- **Noise experiments**—Used to identify the magnitude and directional effect of statistically significant noise factors that will later be used to induce stress in a robustness-optimization experiment

- **Robustness-optimization experiments**—Used to investigate and quantify the interactions between statistically significant control factors and statistically significant noise factors

Statistically significant control factors, along with the interactions between other control factors, are identified during the Design phase of the CDOV process. Statistically significant noise factors are identified during this phase, the Optimization phase, of the CDOV process. The noise diagram is always associated with subsystems and subassemblies, not the system.

Figure 6.9 shows an example of a noise diagram template.

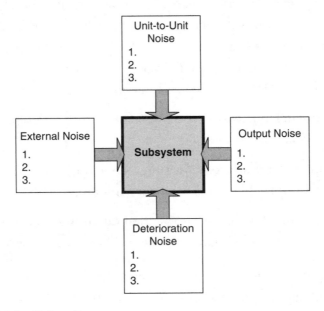

FIGURE 6.9   Noise diagram.

Noises are accounted for at the system level using a *system noise map*. A system noise map illustrates the nature of how subsystems and subassemblies transfer noises across interface boundaries, as well as how noises enter and leave the system from an external perspective. System noise maps show all the noise diagrams as an interconnected network, so they can get quite large for big systems. System noise maps are used to help construct the designed experiments for the system-level stress tests, which come a little later (Phase 3B), after the subsystems and subassemblies have been proven to be robust. System noise maps are often rather frightening things to see. They show just how many noises are running around inside and outside the system, potentially wreaking havoc on the capability of the integrated system to fulfill its various requirements. Failure to construct and look at noise diagrams and system noise maps is failure to look at the design realistically.

Figure 6.10 illustrates an example system noise map template. The arrows signify transmitted noise.

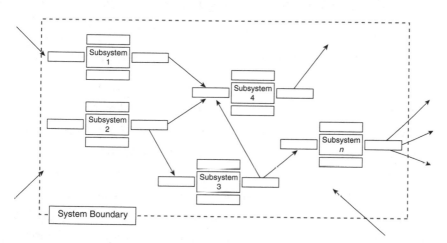

FIGURE 6.10   System noise map.

Conduct Noise Factor Experiments

The subsystem and subassembly noise diagrams are now used to construct a designed experiment to test the statistical significance of the noise factors. The team simply cannot afford to be testing the robustness of CFRs with noise factors that are indistinguishable from random events taking place in the evaluation environment. Noise factors might seem significant in a conference room; this experiment generates the data to find out if they are actually significant. Opinions are often wrong about noise factors—they need to be tested to see if they are truly significant using statistically sound methods of data gathering and analysis.

The noise experiment is often done using a special type of screening experiment called a Plackett-Burman array. The data is analyzed using the analysis of variance technique. The analysis shows which noise factors are statistically significant and which are not. These can also provide data to indicate the magnitude and directional effect of the noise factors (called a *noise vector*) on the mean and standard deviation of the CFRs being measured.

Define Compounded Noises for Robustness Experiments

Because robustness-optimization experiments cost money, it is common to try to design them to be as short and small as possible. One way to keep a robustness stress test economical is to *compound or group* the noise factors into logical groups of two. One grouping sets all the noise factors at the set points that cause the mean of the CFR to get smaller. The other grouping sets up the same statistically significant group of noise factors so that they drive the mean of the CFR high. This approach uniformly stresses the subsystem or subassembly by allowing every combination of control factors to interact with the noise factors. This is accomplished by exposing the treatment combinations of the control factors to the full, aggregated effects of the compounded noise factors.

In this scenario, three control factors exist at two levels and three noise factors exist at two levels. To study every combination of the control factors and noise factors, we need 64 experiments. If we compound the noise factors and stress by noise vector effect, as we have done here, we need only 16 experiments. The key is to use only the compounded effect of statistically significant noise vectors.

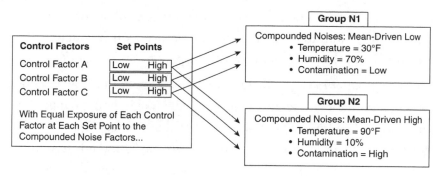

FIGURE 6.11   Compounded noises.

After the control factors have interacted with the compounded noise factors, the data is loaded with the information needed to isolate control factors that leave the design minimally sensitive to the noise factors.

Define Engineering Control Factors for Robustness-Optimization Experiments

Deciding which control factors to place into interaction with the statistically significant noise factors is a matter of engineering science and design principles that are underwritten by the Ideal function equations developed in the Design phase of the CDOV process. The logical choices are the control factors that play a dominant role in the flow, transformation, or state of energy, mass, and controlling signals within the design.

Design for Additivity and Run Designed Experiments

Because the laws of conservation of mass and energy are mathematically defined with additive equations (a form of ideal function), we tend to try to match up control factors and critical functional responses that obey these additive relationships. Setting up a robust design experiment according to this logic is called *designing for additivity*. This is done to bypass unwanted control factor-to-control factor interactions that can arise from measuring quality metrics that possess discontinuous functions with respect to the control factors.

This is an attempt to further our effort to measure fundamental, functional relationships between specifiable engineering parameters and the critical functional responses that they control. Spending the time to set up a robustness experiment this way also tends to require very simple experimental arrays because few control factor-to-control factor interactions end up being evaluated in the robustness evaluations. Recall that the main focus in robust design is to study control factor-to-noise factor interactions.

Analyze Data and Build a Predictive Additive Model

When the designed experiment has been conducted and the CFR data has been collected using the certified measurement system, the signal-to-noise metrics can be calculated. Signal-to-noise transformations are calculated based on the physics and design principles that underlie the design's ideal function. Signal-to-noise metrics are one way of looking at a summary statistic that quantifies, in numeric form, the strength of the interaction between the control and noise factors. Large differences between S/N values for individual control factor set points indicate an interaction with the noise factors. This is a good thing. If a control factor has no interaction with noise factors, robustness is impossible to improve. Gains in robustness come directly from interactions between control and noise factors. The larger the S/N gain, the more insensitive the CFR becomes to variation (providing that you set the control factor nominal value at its most robust set point and then define the appropriate range of tolerances that help constrain the design's robustness even in the presence of unit-to-unit variation).

The units of the S/N metric are calculated and reported in the logarithmic unit called a *decibel* (dB). Decibel units are additive. So as we add up all the individual control factor contributions to robustness, we build an equation known as the predictive additive model (for n number of control factors that actually increase robustness):

$$S/N_{predicted} = S/N_{avg} + (S/N_A - S/N_{avg}) + (S/N_B - S/N_{avg}) + ... + (S/N_n - S/N_{avg})$$

Some control factors simply do not interact significantly with noise factors, so they have no term to include in the predictive additive model.

Run Verification Experiments to Verify Robustness

It is good design practice to conduct an experiment to quantify the subsystem and subassembly baseline S/N values prior to robustness optimization. That way, the new, optimized S/N values can be compared to the baseline S/N values. This helps the team recognize the amount of improvement gained in the optimization process. The gain in the compared S/N values is literal because we have transformed the CFRs from their engineering units into decibels. For every 3dB of gain in robustness, we see the variance of the CFR drop by a factor of 2 (every 6dB of gain is equal to lowering the standard deviation by a factor of 4). The standard deviation is lower when we exploit the interactivity between certain control factor and noise factor combinations. The verification experiment provides the data to assure the team that it has correctly found an additive set of control factors that leave the CFR minimally sensitive to the noise factors.

The final proof of successful robustness optimization resides in the data gathered from confirmation tests. Verification experiments enable the team to confirm that the proper control factors have been identified, with repeatable results in the presence of the statistically significant noise factors. The predicted S/N value must be matched or repeated within a range of tolerance. The tolerance range is calculated using one of several options and is commonly set at approximately +/–2dB to 3dB of the predicted S/N value. If the predicted

S/N and the verification test S/N values are repeatable (usually three to five replicates are tested) within the tolerance zone (particularly within the low side), the additive model is confirmed.

Conduct Response Surface Experiments onCritical Adjustment Parameters

Taguchi's approach to robustness optimization includes a method for identifying which control factors strongly affect the mean performance of the CFR. At least two kinds of control factors are identified during the Optimization phase:

- Control factors that have a strong effect on robustness of the CFRs

- Control factors that have a strong effect on adjusting the mean of the CFRs

Two or more of the control factors that have a significant capability to adjust the mean usually undergo additional testing. These parameters are candidates for designation as critical adjustment parameters. They are the parameters that the assembly and service teams or the customer will use to adjust the mean of specific CFRs onto a desired target. Of course, this has significance in the long-term capability of the design's performance. Recall that Cpk is the measure of long-term performance capability. When we use statistical process control to track the CFR's performance and we find that the mean has shifted off the target, we use the critical adjustment parameters (CAPs) to put the mean back onto the desired target.

Response surface methods are very good at identifying and optimizing CAPs. They are a special class of designed experimental methods that provide very detailed information about the input-output relationships between the mean of the CFR and a small number of critical adjustment parameters. Response surfaces are actually plots of experimental data that resemble contour maps that hikers often use. If two CAPs are being studied, the response surface is in three dimensions and can plot the slope of adjustment sensitivity so the team members can clearly see how to precisely adjust the CAPs to put the CFR exactly where they want it. The RSM approach can be used to adjust the standard deviation in a similar manner, but it is most often used for precision mean adjustment using two or three or more (but rarely more than six) CAPs.

Run Verification Experiments to Verify Tunability and Robustness Parameters for Subsystems and Subassemblies

Now that the control factors have been broken down into two classes, robustizers and mean adjusters, they can all be tested together one last time to prove that the subsystem or subassembly design is both robust and tunable. Taguchi methods have an approach to do this in one step, called the *Dynamic method*. In many cases, it is enough to reach a reasonably optimum (a local optimum) design. When very high tuning precision is required (a global optimum is sought), response surface methods are required. If the design loses its robustness as it is adjusted to a different mean, the team needs to know this as early as possible. In fact, ensuring that this is not a characteristic of the design should take place during the technology-development process. This is where Taguchi's dynamic methods are most powerful in preventing overly sensitive designs. As you can see, we prefer to have relatively high independence between control factors that adjust robustness and those that adjust the mean. This is why we must understand control factor–to–control factor interactions long before we start the Optimization phase of the CDOV process. You have no business conducting Taguchi methods for robust design until you have documented the appropriate deliverables from the tools and best practices within the Design phase. With detailed knowledge of the two classes of interactions (CF x CF interactions identified during the Design phase and CF x NF interactions identified during the Optimization phase), we can efficiently complete an independent optimization of robustness and mean performance. We believe that both Taguchi dynamic robustness evaluations and classical RSM techniques are required in their proper sequence to do the job right—neither is optional or a replacement for the other! If you don't see the wisdom in this, you need to revisit the logic of modeling and robustness in flow of the sequential design of experiments.

Document Critical Functional Parameter Nominal Set Points and CAP and CFR Relationships

With the optimization data gathered and analyzed, the team can document the results in the critical parameter management database. The ideal/transfer functions, now in their additive S/N model form, can be updated. Additional knowledge of control factor designations can be placed in the hierarchical structure of the database for each subsystem and subassembly. The data for which control factors affect CFR robustness and CFR mean performance must be thoroughly documented for the production and sustaining engineering community. When they need to make decisions about cost reduction or design changes, detailed information actually will be available. Production engineering organizations are almost never given enough design information to enable the ongoing support of the design. Critical parameter management corrects this problem.

Develop, Conduct, and Analyze Reliability/Capability Evaluations for Each Subsystem and Subassembly

It is often a waste of time to conduct reliability tests on subsystems and subassemblies that have not yet been optimized for robustness and mean performance. Before the gate review for the Optimization phase, the subsystem and subassemblies should be placed into appropriately designed life testing. The life tests do not have to be completed before the Gate 3A review, but they should be underway. The data from these tests will be heavily reviewed at the Gate 3B review.

Each subsystem and subassembly should also be measured for their CFR design capability performance after robustness optimization is completed. Short-term capability (Cp) evaluation and analysis should be conducted for each critical functional response. This data is to be stored in the critical parameter management database for a capability growth index review at the Gate 3A review.

Gate 3A Readiness

At Gate 3A, we stop and assess the full set of deliverables and the summary data used to arrive at them. The tools and best practices of Phase 3A have been used and their key results can be summarized in a Gate 3A scorecard. A checklist of Phase 3A tools and their deliverables can be reviewed for completeness of results and corrective action in areas of risk.

Phase 3A gate review topics:

- Business/financial case development

- Technical/design development

- Manufacturing/assembly/materials management/supply chain development

- Regulatory, health, safety, environmental and legal development

- Shipping/service/sales/support development

A General List of Phase 3A Tools, Methods, and Best Practices

- Subsystem and subassembly noise diagramming

- System noise mapping

- Measurement system analysis

- Use of screening experiments for noise vector characterization

- Analysis of means (ANOM) data-analysis techniques

- Analysis of variance (ANOVA) data-analysis techniques

- Baseline subsystem and subassembly CFR signal-to-noise characterization

- Taguchi's methods for robust design

- Design for additivity

- Full or fractional factorial experiments for robustness characterization

- Generation of the additive S/N model

- Response surface methods

- Design capability studies

- Critical parameter management

- Reliability analysis

- Life testing (normal, HALT, and HAST)

Table 6.3 integrates the Optimize Phase 3A phase requirements, deliverables, tasks, and enabling tools.

TABLE 6.3 Tools, Tasks, Deliverables-to-Requirements Integration Table

Requirement	Deliverable(s)	Task(s)	Tool(s)
Sublevel designs that are robust and tunable	Documented robustness and tunability of each sub-level design (at functional assembly level and higher)	Review and finalize the critical functional responses for the subsystems and subassemblies.	Subsystem and subassembly noise diagramming
		Develop subsystem noise diagrams and system noise map.	System noise mapping
		Conduct noise factor experiments.	Measurement system analysis
		Define compounded noises for robustness experiments.	Use of screening experiments for noise vector characterization
		Define engineering control factors for robustness-optimization experiments.	Analysis of means (ANOM) data-analysis techniques
			Analysis of variance (ANOVA) data-analysis techniques
		Design for additivity and run designed experiments or simulations.	Baseline subsystem and subassembly CFR signal-to-noise characterization
		Run verification experiments to verify robustness.	Taguchi's methods for robust design
		Analyze data and build predictive additive model.	Design for additivity
			Use of full or fractional factorial experiments for robustness characterization
		Conduct response surface experiments on critical adjustment parameters.	Generation of the additive S/N model
		Run verification experiments to verify tunability and robustness parameters for subsystems and subassemblies.	Response surface methods
			Design capability studies

continues

TABLE 6.3   Tools, Tasks, Deliverables-to-Requirements Integration Table (continued)

Requirement	Deliverable(s)	Task(s)	Tool(s)
Sublevel design reliability growth	Documented reliability growth of all sublevel designs (under nominal and stressful conditions)	Develop, conduct, and analyze reliability/capability evaluations for each subsystem and subassembly.	Reliability analysis Life testing (normal, HALT, and HAST)
Sublevel design risk assessments	Documented risk assessment of the sublevel designs	Conduct DFMEA on sublevel designs.	DFMEA
Critical parameter management database	Documented sublevel critical parameters	Document critical functional parameter nominal set points, CAP and CFR relationships.	Design capability studies Critical parameter management

Phase 3B: System Integration, Nominal Performance Evaluation, Robustness Optimization, and Initial System Reliability Assessment

The Optimize tasks for Phase 3B include the following:

1. Integrate sublevel designs into system test units.

2. Verify capability of the system-wide data-acquisition system.

3. Conduct nominal performance evaluation on the system.

4. Conduct system robustness stress tests.

5. Refine subsystem set points to balance system performance.

6. Conduct initial system-reliability assessment.

7. Verify system readiness for final product design capability development and assessment.

8. Document system risk profile.

9. Generate Verify phase project plans.

Integrate Sublevel Designs into System Test Units

When the subsystems and subassemblies have fulfilled their 3A gate requirements, they are safe to proceed into the Phase 3B integration phase of the CDOV process. It is extremely risky to allow nonrobust subsystems with poor adjustment capability to be integrated into the system. Many names are used to define the product as it is integrated into physical hardware, firmware, and software. Some refer to it as the system breadboard; others call the embryonic product the system integration fixture, test rig, or unit. Whatever you call it, it represents the first completely assembled set of design elements that are attempting to fulfill the system requirements.

It is wise to conduct a fairly rigorous program of inspection on all the components that make up the subsystems and subassemblies that are being integrated into the system. If this is not done, the system-integration team has no idea what it has built. Only after testing begins do the anomalies begin to surface. You must prevent the whole set of errors, mistakes, and false starts that come with "bad parts" getting built into the integration units. A great deal of time is needed to sort out these problems if they are buried in the system from the onset of testing. The famous quote of "Measure twice, cut once" certainly applies here. Design cycle-time will be reduced if you don't have to go through a whole cycle of finding mistakes due to these kinds of issues.

Verify Capability of the System-Wide Data-Acquisition System

As the subsystems and subassemblies are integrated into the system test units, the transducers and data-acquisition wiring should be installed, tested, and certified. The system-integration team will be measuring a select set of subsystem and subassembly critical functional responses and all the system-level critical functional responses.

The critical adjustment parameters must be carefully measured for their effect on their associated CFR variables. This is a major focus of the data-acquisition system. The critical adjustment parameters and the critical functional responses they tune must be thoroughly measured and their sensitivity relationships documented in the CPM database. Measurement system analysis must be thoroughly applied before system testing.

Many subassembly-to-subsystem-to-system transfer functions will be derived from the data taken during the system-integration tests. Corrupt data destroys the integrity of these critical paths of transferred functional performance and variability.

Conduct Nominal Performance Evaluation on the System

After the system is integrated and the data-acquisition system is certified, the system-integration test team is ready to conduct nominal system performance testing. If the system-level CFRs cannot be measured because the system fails to function or the system simply won't run well enough to gather a nominal baseline of performance, there is no sense going on with system stress testing. The primary goal of this first evaluation of integrated system performance is to characterize the initial Cp values for all the system CFRs and the selected subsystem and subassembly CFRs. Additional goals are to identify and document unforeseen problems, mistakes, and errors of omission and their corrective actions across the newly integrated system. The work done in the previous phases was conducted to prevent as many problems, mistakes, errors, and omissions as possible—but some will still be encountered.

Conduct System-Robustness Stress Tests

Depending on the nature and severity of the problems stemming from the nominal performance tests, the system-integration team can proceed with a series of stress tests. The system stress tests are typically designed experiments that evaluate all the system CFRs as a well-defined set of critical functional parameters (subsystem and subassembly CFRs) and critical-to-function specifications (component CTF set points) that are intentionally set at all or part of their tolerance limits. The system stress tests also include the inducement of various levels of compounded external and deterioration noise factors.

This form of testing is expected to reveal sensitivities and even outright failures within and across the integrated subsystems and subassemblies. The magnitude and directional effect of the noise factors on the system CFRs is of great importance relative to how the subsystem, subassembly, component, and manufacturing process nominal and tolerance set points will be changed and ultimately balanced to provide the best system performance possible.

Refine Subsystem Set Points to Balance System Performance

Calculating the Cp values for each system-level CFR is recommended during this phase. These Cp values will be relatively low, as you might expect. It is quite helpful to know, at this early phase in the commercialization process, what the nominal and stress-case Cp values are across the system. It is not unusual to take a number of corrective actions and to move a number of subsystem, subassembly, and component set points to help balance and improve system CFR performance. This is one reason why we need robust and tunable designs. Any number of post-corrective action system stress tests can be run as needed to verify performance improvements. The usual number is two: one initial stress test and one just before exiting Phase 3B. So a minimum of two rounds of stress testing are recommended for your planning and scheduling consideration. Often the first round of stress testing is conducted at the tolerance limits of the critical parameters selected for the test. The second round of stress testing, after design-balancing analysis and corrective action is completed, is less aggressive. It is quite reasonable to set the critical parameters at a third of their rebalanced tolerance limits. It is not unusual to go through an initial round of strategic tolerance tightening at the subsystem level during this phase of performance testing and balancing.

Conduct Initial System Reliability Assessment

With the nominal and stress testing completed and corrective actions taken, it is now worth the effort and expense to conduct a formal system-reliability evaluation. The system is set to nominal conditions with all components, subassemblies, and subsystems inspected and adjusted to produce the best performance possible at this phase of product commercialization.

Subsystem, subassembly, and component life-tests should be ongoing. Environmental stress screening (ESS), highly accelerated life-testing (HALT), and highly accelerated stress testing (HAST) should be producing data. Failure modes should be emerging and corrective action should be underway to increase reliability. DFMEAs should be updated.

Verify System Readiness for Final Product Design Capability Development and Assessment

After conducting all the nominal and stress tests and getting a first look at reliability test data, the system has been evaluated to a point at which a recommendation can be made relative to the system's "worthiness" of further investment in the next phase (4A and 4B). If the system is too sensitive to noise factors, corrective action to remove the sensitivities will have to be done or the program will have to be cancelled or postponed.

The critical parameter management database is updated at this point. All the new sensitivity relationships and changes to the ideal/transfer functions need to be included. The latest system, sub-system, and subassembly CFR Cp performance data also should be added for the Gate 3B review.

Document System Risk Profile

Using the critical parameter data and the results from the reliability assessment test, document all significant risks that have accrued over this phase of commercialization.

Generate Verify Phase Project Plan

Generate the detailed project plans for the next phase of commercialization. Apply Monte Carlo simulations to the critical path of tasks and conduct a project cycle-time FMEA. Document a risk-mitigation plan to prove that you are capable of preventing problems over the next phase of work.

Gate 3B Readiness

At Gate 3B, we stop and assess the full set of deliverables and the summary data used to arrive at them. The tools and best practices of Phase 3B have been used and their key results can be summarized in a Gate 3B scorecard. A checklist of Phase 3B tools and their deliverables can be reviewed for completeness of results and corrective action in areas of risk.

Phase 3B Gate Review Topics

- Business/financial case development
- Technical/design development
- Manufacturing/assembly/materials management/supply chain development
- Regulatory, health, safety, environmental, and legal development
- Shipping/service/sales/support development

A General List of Phase 3B Tools, Methods, and Best Practices

- Measurement system analysis
- System noise mapping
- Nominal system CFR design Cp studies
- Stress-case system CFR design Cp studies

- System-subsystem-subassembly transfer function development

- System-level sensitivity analysis

- Design of experiments (screening and modeling)

- ANOVA

- Regression

- Analytical tolerance analysis

- Empirical tolerance analysis

- Statistical process control of design CFR performance

- Reliability assessment (Normal, HALT, and HAST)

- Critical parameter management

- Monte Carlo simulation

- Project cycle-time FMEA

Table 6.4 integrates the Optimize Phase 3B requirements, deliverables, tasks, and enabling tools.

Verify Phase: Verification of Final Product Design, Production Processes, and Service Capability

Product Commercialization

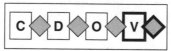

Phase 4 contains two major parts:

- Phase 4A focuses on the capability of the product design functional performance.

- Phase 4B focuses on the capability of production assembly and manufacturing processes within the business, as well as the extended supply chain and service organization.

TABLE 6.4 Tools, Tasks, Deliverables–to–Requirements Integration Table

Requirement	Deliverable(s)	Task(s)	Tool(s)
System that is robust and tunable	Documented robustness and tunability of the integrated system	Integrate sublevel designs into system test units.	Measurement system analysis
		Verify capability of the system-wide data-acquisition system.	System noise mapping
		Conduct nominal performance evaluation on the system.	Nominal system CFR design Cp studies
		Conduct system robustness stress tests.	Stress-case system CFR design Cp studies
		Refine subsystem set points to balance system performance.	System-subsystem-subassembly transfer function development
			System-level sensitivity analysis
			Design of experiments (screening and modeling)
			ANOVA
			Regression
			Analytical tolerance analysis
			Empirical tolerance analysis
			Statistical process control of design CFR performance
System design reliability growth	Documented reliability growth of the integrated system (under nominal and stressful conditions)	Conduct initial system reliability assessment.	Reliability assessment (normal, HALT, and HAST)
		Verify system readiness for final product design capability development and assessment.	

System risk profiles	Documented risk profile of the system	Document system risk profile.	DFMEA
Critical parameter management database	Documented system-level critical parameters	Update CPM database.	Capability studies Critical parameter management
Verify phase project plans	Documented project plans for the Verify phase	Generate Verify phase project plans.	Monte Carlo simulation Project cycle-time FMEA

The Verify gate requirements include the following:

- Sublevel design capabilities

- Sublevel design reliability growth

- Sublevel design risk assessment

- System capability

- System design reliability growth

- System risk profiles

- Manufacturing and assembly capability

- Critical parameter management database

- Launch plans

The Verify gate deliverables include these:

- Documented capability of each sublevel design, with and without stress (at functional assembly level and higher)

- Documented reliability growth of all sublevel designs (under nominal and stressful conditions)

- Documented reliability growth of the integrated system (under nominal and stressful conditions)

- Documented capability of the integrated system, with and without stress

- Documented risk assessment of the sublevel designs

- Documented risk profile of the integrated system

- Documented sublevel critical parameters

- Documented system-level critical parameters

- Documented corrective action plans

- Documented launch plans for the Launch phase of LMAD

The Verify tasks for Phase 4A include these:

1. Conduct final tolerance design on components, subassemblies, and subsystems.

2. Place all critical-to-function components and critical functional responses under statistical process control in supply chain and assembly operations.

3. Build product design verification units using production parts.

4. Evaluate system performance under nominal conditions.

5. Complete corrective actions on problems.

6. Evaluate system performance and reliability.

7. Verify that the product design meets all requirements.

8. Develop transfer plan for the critical parameter database for production, supply chain, and service organizations.

Conduct Final Tolerance Design on Components and Subsystems

In Phase 3B, the subsystems and subassemblies passed through their initial system-integration evaluations. Numerous adjustments and refinements at the subsystem, subassembly, and component levels were made. These were mainly shifts in the nominal set points of the CFRs and CTF specifications to balance overall system CFR performance under both nominal and stressed conditions. Initial tolerances were defined during the Design phase of the CDOV process. Numerous critical parameter (subsystem and subassembly CFRs) and specification (component) tolerances were exercised during the system stress tests. Some of these critical tolerances were adjusted to help mature and improve the system performance to pass Gate 3B requirements.

Now we are ready to develop the final tolerances on all manufacturing process set points, components, subassemblies, and subsystems. This is where critical parameter management comes into full use. The entire network of nominal and tolerance set points must be finalized and verified so that they fulfill all the requirements. All other nominal and tolerance set points that are of general importance also must be finalized.

The methods of analytical and empirical tolerance design, along with design and manufacturing capability studies, can be used extensively to verify the tolerances that will constrain the right amount of unit-to-unit noise. The unit-to-unit noise must be held to a level that enables all the requirements throughout the integrated system to be fulfilled to the best extent possible. This becomes an economic issue because the tightening of tolerances and improvement of bulk or surface material properties always increases component costs.

Analytical tolerancing is typically conducted on a computer using math (ideal/transfer functions applied within and across a design boundaries) and geometric models. Empirical tolerancing is conducted in a design laboratory or production facility using designed experiments to evaluate candidate tolerances on real components, subassemblies, and subsystems. In both cases, numerous tolerances are set at various high and low levels, and then the model or prototype exercises the effect of the tolerance swings within the ideal/transfer function being evaluated. The resulting variability exhibits its effects in the measured or calculated CFRs.

Place All CTF Components and CFRs under SPC in Supply Chain and Assembly

After all the final tolerances for the components, subassembly adjustments, and subsystem adjustments have been defined, the ones that are critical-to-function are documented and placed under statistical process control. The critical parameters can be measured when components are assembled and adjusted during the assembly process. Critical specifications are measured during component production as mass, energy, and information and are transformed within a production process. Statistical process-control methods are used to determine when variation and mean performance have left a state of random variation and have entered a state of assignable cause variation. Engineering analysis, multivari studies, capability studies, and designed experimentation can help identify and remove assignable causes of variation.

This approach to variability reduction in the mean and standard deviation of critical parameters and specifications is not cheap. That is why we use CPM: It helps the team spend money only on those design and production process elements that are critical to the functional requirements of the product.

Build Product Design Certification Units Using Production Parts

When the all the elements of the system design are toleranced and the assembly and production process are under statistical control, the first sets of production parts can be ordered, inspected, and assembled. The production teams construct the first production systems with the design teams present to document any problems.

The list of critical functional responses that will be measured during the assembly process at the subassembly, subsystem, and system levels is prepared at this point. Not all the CFRs within the system that were measured during the CDOV process will continue to be measured during steady-state production. Only a small subset of the CFRs will need to go forward as part of the manufacturing critical parameter management program for Six Sigma. Many will be measured as CTF specifications on components out in the supply chain, a few will be measured as CFRs, as subassemblies and subsystems are assembled and tested before they are integrated into the final product assembly. The remaining CFRs will be measured at the system or completed product level usually just before shipping or during service processes.

Evaluate System Performance under Nominal Conditions

When the system is built and all CFR measurement transducers are in place, a repeat of the nominal system performance test can be conducted. The Cp of the system CFRs and the production subsystem and subassembly CFRs is measured and calculated. The aggregated affect of these Cp values is reported in a metric called the *capability growth index*.

Evaluate System Performance under Stress Conditions

The new, final tolerances for critical-to-function components and the critical assembly and service adjustments are typically set at one-third (for Three Sigma designs) to one-sixth (for Six Sigma designs) of their limits for testing purposes. In some cases, this roughly emulates what is known in statistical tolerancing methods as a root sum of squares evaluation. It is more statistically likely that the variability that will actually accumulate within and across the system will not be at worst-case limits, but rather some lesser amount of consumption of the tolerance limits. From experience and from statistical principles, we have found it reasonable over the years to stress the performance of the critical response variables with components and adjustments at about one-third to one-sixth of their actual tolerances. The production CFRs are measured during these stress tests with the added stress due to applications of external and deterioration noise factors from the system noise map.

Once again, the Cp of the system CFRs and the production subsystem and subassembly CFRs are measured and calculated. The aggregated affect of these stressed Cp values is reported in the same capability growth index (CGI) format. The measured performance becomes the data that is used in documenting final product design specifications that represent the teams' best attempt to fulfill the system, subsystem, subassembly, and component requirements.

Complete Corrective Actions on Problems

As is always the case, after nominal and stress testing, problems will surface and must be corrected. The key is to prevent as many of them as possible by using the right tools and best practices at the right time within each phase of the CDOV process. Problems at this point in the process should be relatively minor.

Evaluate System Performance and Reliability

As you can see, this approach postpones traditional reliability assessments until corrective actions are largely completed after nominal and stress testing is done within each major phase. There is no sense assessing reliability when you know full well you don't have the design elements mature enough to realistically assume that you are anywhere near the goal you seek to confirm. Developing and evaluating ideal functions, transfer functions, noise vectors, additive S/N models, tolerance sensitivity stack-up models, and capability growth models are all steady and progressive actions that will develop reliability. Reliability tests never grow reliability; they just tell you how much you do or don't have. They are very important when conducted at the right time and for the right reasons. They confirm that you are attaining your goals, but should not be done too early. The rule is, don't consume any resources on predicting or testing reliability that really could be better utilized in the actions of developing reliability. Many of the tools and metrics of DFSS are there because they are good at creating reliability long before you can actually measure it.

System-reliability evaluations typically demonstrate the performance of the functions of the system in terms of mean time to failure or mean time to service. The underlying matters of importance in reliability testing are as follows:

- Time

- Failures

Customers can easily sense or measure both of these. The product-development teams must not ignore these metrics. In DFSS, we must keep two sets of books to track the fulfillment of all the product requirements. The *book of requirements* that customers worry about is typically at the system level and is rarely stated in engineering units. To fully satisfy these quality requirements that customers measure against, we must have a set of books that dig deeper, down through the hierarchy of the product and production systems' architectures. This *book of translated requirements* is loaded with fundamental engineering metrics that can provide the preventative, forward-looking metrics that ensure that the functions can indeed be replicated, over the required lengths of time that will fulfill the VOC requirements. Our metrics have to be good enough to measure and make us aware of impending failures.

Verify That Product Design Meets All Requirements

The ability to verify the system design requires two things:

- A clear set of design requirements at the system, subsystem, subassembly, component, and manufacturing process levels

- A capable set of design performance metrics for the critical functional responses at the system, subsystem, and subassembly levels, and a capable set of metrics for the critical-to-function specifications at the component and manufacturing process levels

The construction and use of this integrated system of requirements and metrics is called critical parameter management. It will be much easier to rapidly determine the status of capability to fulfill form, fit, and function requirements if a very disciplined approach is taken from the first day of the CDOV process to track these critical relationships. Critical parameter management possesses a scorecard that accumulates Cp performance growth across the subsystems and at the system level. This scorecard is populated with Cp values from all the subsystem critical functional responses and a rolled-up summary capability growth value called the capability growth index (CGI). It is somewhat analogous to rolled throughput yield (RTY) that is used in production operations and transactional forms of Six Sigma. The CGI is based on CFRs, while RTY is typically based on defects per million opportunities or defects per unit. The CGI is the management team's eyes into the "soul" of the product from a functional perspective. When the CGI scorecard is at its required level of Cp growth, the quality is inherent in the certified product design.

Develop Transfer Plan for the Critical Parameter Database for Production, Supply Chain, and Service Organizations

The final action for the technical team is to develop a clear plan for transferring the appropriate critical parameter data to the production, supply chain, and service organizations. This database is essential for these teams to sustain the capability of the product out in the launch and product lifecycle environment. Only the critical parameter data that adds value to the ongoing support of the product should be documented and transferred to each of these organizations. Part of the lean mentality is to provide only information that is absolutely necessary.

Gate 4A Readiness

At Gate 4A, we stop and assess the full set of deliverables and the summary data used to arrive at them. The tools and best practices of Phase 4A have been used and their key results can be summarized in a Gate 4A scorecard. A checklist of Phase 4A tools and their deliverables can be reviewed for completeness of results and corrective action in areas of risk.

Phase 4A Gate Review Topics

- Business/financial case development

- Technical/design development

- Manufacturing/assembly/materials management/supply chain development

- Regulatory, health, safety, environmental, and legal development

- Shipping/service/sales/support development

A General List of Phase 4A Tools and Best Practices

- Measurement system analysis

- System noise mapping

- Analytical tolerance design

  - Worst-case analysis

  - Root sum of squares analysis

  - Six Sigma analysis

  - Monte Carlo simulation

- Empirical tolerance design

  - System-level sensitivity analysis

- Design of experiments

  - ANOVA

  - Regression

- Multivari studies

- Design capability studies

  - Nominal system CFR design Cp studies

  - Stress-case system CFR design Cp studies

- Reliability assessment (normal, HALT, and HAST)

- Competitive benchmarking studies

- Statistical process control of design CFR performance

- Critical parameter management (with capability growth index)

Table 6.5 integrates the Verify Phase 4A requirements, deliverables, tasks, and enabling tools.

Phase 4B: Capability of Production, Assembly, and Manufacturing Processes within the Business as well as the Extended Supply Chain and Service Organization

The Verify tasks for Phase 4B include the following:

1. Build initial production units using inspected production parts.

2. Assess capability of all CFRs and CTFs in production and assembly processes.

3. Assess capability of all product level and sublevel level CFRs during assembly.

4. Assess reliability of production units.

TABLE 6.5 Tools, Tasks, Deliverables-to-Requirements Integration Table

Requirement	Deliverable(s)	Task(s)	Tool(s)
Sublevel design capabilities	Documented capability of each sublevel design, with and without stress (at functional assembly level and higher)	Conduct final tolerance design on components, subassemblies, and subsystems. Place all critical-to-function components and critical functional responses under statistical process control in supply chain and assembly operations.	Measurement system analysis Analytical tolerance design Worst-case analysis Root sum of squares analysis Six Sigma analysis Monte Carlo simulation Empirical tolerance design Design of experiments ANOVA Regression Multivari studies Design capability studies
Sublevel design reliability growth	Documented reliability growth of all sublevel designs (under nominal and stressful conditions)	Evaluate system performance and reliability.	Reliability assessment (normal, HALT, and HAST)
Sublevel design risk assessment	Documented risk assessment of the sublevel designs	Conduct DFMEA.	DFMEA

System capability	Documented capability of the integrated system, with and without stress	Build product design verification units using production parts. Evaluate system performance under nominal conditions. Complete corrective actions on problems.	Measurement system analysis System noise mapping Analytical tolerance design Worst-case analysis Root sum of squares analysis Six Sigma analysis Monte Carlo simulation Empirical tolerance design System-level sensitivity analysis Design of experiments ANOVA Regression Multivari studies Design capability studies Nominal system CFR design Cp studies Stress-case system CFR design Cp studies
System design reliability growth	Documented reliability growth of the integrated system (under nominal and stressful conditions)	Evaluate system performance and reliability.	Reliability assessment (normal, HALT, and HAST)
System risk profile	Documented risk profile of the integrated system	Conduct DFMEA.	DFMEA

continues

TABLE 6.5 Tools, Tasks, Deliverables-to-Requirements Integration Table (continued)

Requirement	Deliverable(s)	Task(s)	Tool(s)
Critical parameter management database	Documented sublevel critical parameters Documented system-level critical parameters Documented corrective action plans	Develop transfer plan for the critical parameter database for production, supply chain, and service organizations. Verify that the product design meets all requirements.	Competitive benchmarking studies Statistical process control of design CFR performance Critical parameter management (with capability growth index)
Launch plans	Documented launch plans for the Launch phase of LMAD	Develop transfer plan for the critical parameter database for production, supply chain, and service organizations.	Statistical process control of design CFR performance Critical parameter management (with capability growth index)

5. Verify that all requirements are being met across assembly processes.

6. Verify that all requirements are being met across production processes.

7. Verify that all service requirements are being met with service/support processes.

8. Verify that product, production, assembly, and service/support processes are ready for launch.

Build Initial Production Units Using Inspected Production Parts

Now that the organization has completed the certification of the final product design specifications, it is time to order actual production parts and build the initial production units that will bear revenue. This is not a serial process. Many parts, tools, and subassemblies will have long lead times. A good deal of judgment must be exercised to balance the risk of early ordering based on assumed certification versus waiting until the data is all in before ordering the "hard" tools and components.

Many design organizations get their part orders way out of "phase" with their performance and capability data. The designers keep designing before the testers finish testing! One way to remedy this is to make the designer and tester the same person. At times it is good to have your people so busy doing one thing they can't get anything else done! The real problems come when this phenomenon occurs habitually from the beginning of a new product-development program. By the time the final production units are being built, virtually no one knows if or how well they will actually work.

The production parts need to be produced from and inspected against final, certified design specifications. A product design specification is the actual nominal and tolerance values that the design actually can provide. The motivating forces behind the product design specifications are the system, subsystem, subassembly, component, and manufacturing process requirements. If the parts are not inspected during this earliest production build, you run the risk of not catching a number of human errors and mistakes that, once corrected, are likely to not be a problem ever again. If these mistake-proofing inspections do not occur, it is quite possible, in any sufficiently complex product, for these problems to go on for extended periods of time, particularly if the problem is intermittent.

Assess Capability of All CFRs and CTFs in Production and Assembly Processes

A thorough assessment of the critical parameter management metrics that are transferring into steady-state production must be conducted at this point. This body of metrics is a relatively small, very carefully selected group of CFRs and CTF specifications from the design CPM database. The production CPM database is a scaled-down version of the design CPM database. The design CPM database still must be available to the production organization for the life of the product, but the specific production-oriented CFRs and CTF specifications will be the leading indicators that show that the manufacturing processes and the assembled parts coming from them are capable and under statistical control, ensuring that they are safe to use in steady-state production.

Assess Capability of All Product-Level and Sublevel CFRs During Assembly

Within the actual product will exist a few CFRs that the assembly team will test as the subassemblies and subsystems are constructed. When these are assembled into the product, the system-level CFRs can be evaluated. No need exists for additional, special, or long-term "burn in" style testing when critical parameter management is systematically used across the phases and gates of the CDOV process.

The selection of these CFRs works somewhat analogously to the way your physician conducts your annual physical. Neither you nor your doctor can afford to test everything that is important to your health. The few critical parameters that are the leading indicators of your health can efficiently and economically be evaluated. If you pass these critical checks, out the door you go. This is essentially what must be done with your products. Design in the quality, document and check the critical parameters, and get the product to the customers so that they can be satisfied on time and your shareholders will see a great return on their investment in the business.

Assess Reliability of Production Units

An appropriately small number of the production units should be randomly selected for a complete reliability evaluation. This is done to complete the competitive benchmarking process and to ensure that there are no lingering surprises in the initial production lot of system units. The data from this evaluation can be used as a baseline for your next product-development program.

When you are past the initial ramp-up of your production process, the data you need for reliability should come from the base of delivered products. A strong customer service and support process with a focused effort to establish data acquisition and transmission back to the design organization from a few strategic customers on CFRs and general reliability is essential. This is another way the critical parameter management process is extended into the steady-state use environment.

Verify That All Requirements Are Being Met across Assembly Processes

Just as you established critical parameter management metrics into the supply chain to quantify the capability of incoming component CTF specifications, you should establish a series of CFR and CTF specification checks as the assembly process is conducted. Here we see a portion of the computer-aided data-acquisition systems that were used in design transferred to the assembly team for their continued use during steady-state production.

Verify That All Requirements Are Being Met across Production Processes

A comprehensive roll-up of all critical parameter management data in the network of production processes must be gathered to demonstrate that all the critical manufacturing processes, components, assembled subassemblies, and subsystems are on target with low variability (Cpk measured). The ultimate requirements that matter the most are at the system or completed product level. All system-level CFRs should be found capable on a routine basis. Assignable causes of variation should be identified using statistical process control. They should be removed under the scrutiny and balance of the CPM database.

Verify That All Service Requirements Are Being Met with Service/Support Processes

All CFRs with critical adjustment parameters that the service community will use should be certified for adjustability. The service engineering community will then be able to play its role in sustaining the Cpk values for the system and subsystem CFRs that they are responsible for.

The CPM database should be in the hands of the service engineering teams for diagnostic and maintenance purposes. One of the biggest payoffs in conducting CPM is the detailed technical performance information that can be passed on to the service organization. They can readily use the relational ideal/transfer function information resident in the CPM database to track down root causes of problems and take very specific corrective action. This greatly increases the "fix it right the first time" expectation that modern customers possess.

Verify That Product, Production, Assembly, and Service/Support Processes Are Ready for Launch

As a final Gate 4B deliverable, the entire launch-readiness status, based on the capabilities that can be tracked in the CPM database, should be prepared for review. All the capability data across the organizations that are responsible for verifying launch readiness can be summarized and used to underwrite the integrity of the decision to launch or to postpone for corrective action.

Gate 4B Readiness

At Gate 4B, we stop and assess the final, full set of deliverables and the summary data used to arrive at them. The tools and best practices of Phase 4B have been used and their key results can be summarized in a Gate 4B scorecard. A checklist of Phase 4B tools and their deliverables can be reviewed for completeness of results and corrective action in areas of risk.

Phase 4B Gate Review Topics

- Business/financial case development
- Technical/design development
- Manufacturing/assembly/materials management/supply chain development
- Regulatory, health, safety, environmental, and legal development
- Shipping/service/sales/support development

A General List of Phase 4B Tools, Methods, and Best Practices

- Measurement system analysis
- Analytical tolerance design
  - Worst-case analysis
  - Root sum of squares analysis
  - Six Sigma analysis
  - Monte Carlo simulation
- Empirical tolerance design
  - System-level sensitivity analysis
- Design of experiments
  - ANOVA
  - Regression
- Multivari studies
- Design capability studies
  - Nominal system CFR design Cp studies
- Reliability assessment (normal, HALT, and HAST)
- Competitive benchmarking studies
- Statistical process control of design CFR performance
- Critical parameter management (with capability growth index)

Table 6.6 integrates the Verify Phase 4B requirements, deliverables, tasks, and enabling tools.

TABLE 6.6   Tools, Tasks, Deliverables-to-Requirements Integration Table

Requirement	Deliverable(s)	Task(s)	Tool(s)
Manufacturing and assembly capability System capability	Documented capability of the integrated production system, with and without stress Documented risk assessment of the sublevel production and assembly processes Documented risk profile of the integrated production and supply chain system Documented critical parameters for production and assembly processes Documented corrective-action plans Documented launch plans for the Launch phase of LMAD	Build initial production units using inspected production parts. Assess capability of all CFRs and CTFs in production and assembly processes. Assess capability of all product-level and sublevel CFRs during assembly. Verify that all requirements are being met across assembly processes. Verify that all requirements are being met across production processes.	Measurement system analysis Analytical tolerance design Worst-case analysis Root sum of squares analysis Six Sigma analysis Monte Carlo simulation Empirical tolerance design System-level sensitivity analysis Design of experiments ANOVA Regression Multivari studies Design capability studies Nominal system CFR design Cp studies Competitive benchmarking studies Statistical process control of design CFR performance

continues

TABLE 6.6 Tools, Tasks, Deliverables–to–Requirements Integration Table (continued)

Requirement	Deliverable(s)	Task(s)	Tool(s)
System design reliability growth	Documented reliability performance growth	Assess reliability of production units.	Reliability assessment (normal, HALT, and HAST)
Critical parameter management database	Documented product and production/assembly/service process critical parameters	Verify that all production and service requirements are being met with service/support processes.	Critical parameter management (with capability growth index)
Launch plans	Documented launch plans for the Launch phase of LMAD	Verify that product, production, assembly, and service/support processes are ready for launch.	Critical parameter management (with capability growth index)

TABLE 6.7 Gate Deliverable Review Scorecard

1	2	3	4	5	6
Gate Deliverable	Grand Avg. Tool Score	% Task Completion	Results vs. Requirement	Risk Color Code(R-Y-G)	Gate Requirement

TABLE 6.8 Task Scorecard

1	2	3	4	5	6
Gate Deliverable	Grand Avg. Tool Score	% Task Completion	Task Results vs. Required Deliverable	Risk Color Code (R-Y-G)	Gate Deliverable

TABLE 6.9 Tool Scorecard

1	2	3	4	5	6
Tool	Quality of Tool Use	Data Integrity	Results vs. Required Task Deliverable	Specific Tool Score	Task & Deliverable

7

FAST TRACK
COMMERCIALIZATION

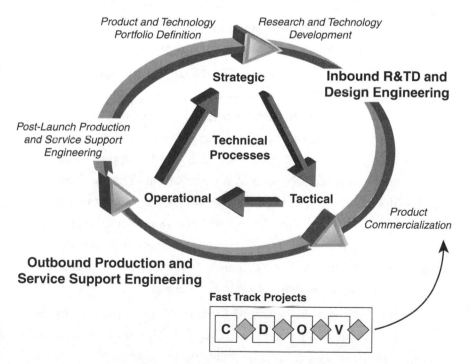

High-risk, high-reward, rapid commercialization: proceed with caution!

Six Sigma Applications for Fast Track Commercialization Projects

Sometimes an opportunity arises to really cash in on a set of circumstances that make it worth the risk to push a project through your phase-gate process at a very high rate. You might call this "rushing" a project. Most projects should be "hurried" along and not rushed. When we use the word *lean* in this context, we are referring to a project that is hurried along through the proper design of a balanced flow of value-adding tasks that focus on fulfilling requirements that are directly tied to recently gathered customer needs.

A Fast Track Project is really beyond what you would call "lean." The main characteristic of a Fast Track Project is that some value-adding work will not get done fully and completely—and some not at all—until a post-launch DMAIC Six Sigma Team completes these tasks. We are making the assumption, and we could be wrong, that you already have put in place some form of Six Sigma competence in the DMAIC problem-solving process tradition.

Some Fast Track Projects will violate other lean principles because they will be driven not by direct customer data, as such, but rather by a new breakthrough in technology that drives such value that customers will simply be blown away by what the product offers. These are cases in which customers did not know they wanted or could use such a paradigm-breaking innovation. This is "lead user" turf. When an innovation breaks down old paradigms and you want to absolutely dominate the market for this new opportunity, justification can be made for a Fast Track Project. This chapter shares our considered opinion on how to do this and cause the least amount of damage to your brand and reputation in the market.

This is high-risk, high-reward territory. It is rational to do some projects this way—but not many. If your business has developed a habit in which you have to rush all new products to market, you need to make a significant effort to control how much of this you actually do. Your product portfolio renewal process needs to be structured to get a better balance of risk profiles across your growth projects (see Chapter 4, "Strategic Product and Technology Portfolio Renewal Process"). We get asked about the design of tool-task-deliverable groups that one can use to minimize risk while a project is put on a fast track to launch. We explain how to use a combination of DMAIC methods in the post-launch environment and DFSS during CDOV phases to conduct a Fast Track Project. You should use the IDEA (see Chapter 4) and IIDOV (see Chapter 5, "Strategic Research and Technology Development Process") processes to ensure that you have the right environment and data to justify a Fast Track Commercialization Project.

DMAIC Six Sigma Project Capability to Support a Fast Track Project

In part, our answer to this Fast Track issue is grounded in your expertise in classic DMAIC Six Sigma and how fast and well you can apply DMAIC tools to clean up unfinished tasks in a post-launch context. If you are really good at DMAIC Six Sigma projects and have the dedicated resources to apply them, you have what we consider a foundation of skills to justify accepting the risk of doing Fast Track Projects.

Core DMAIC Six Sigma competencies include these:

- A hierarchy of master black belts, black belts, and green belts that work well together as a team. Do you have enough of them to conduct unfinished design tasks whose problems will be fully documented in your FRACAS (failure report and corrective action system) or problem report database? You might choose to not do some tasks and happily get away with it. More likely, you will trim tasks and get clobbered with problems—some predicable and others complete surprises. These DMAIC experts should be told exactly what tasks and tools you will not have time to apply so they can get ready to do it after the launch is executed "prematurely by design." Remember, a compounded effect occurs because these people will have to unravel the interactive sensitivities and variation transmissions that your design team rushed past. Think "pit crew" or SWAT team levels of competence and teamwork. These people will get a bit of a mess handed to them. Their project will have to deal with two key areas: to finish tasks that are known to be "not done by design," and to conduct tasks that are surprises, because things that were not anticipated actually happened and now must be dealt with. The more you can document what wasn't done, the better chance the experts will have of quickly diagnosing the shortfalls and determining what tasks and tools are needed to close the gaps.

- A track record of successfully completed projects that have fixed problems similar to the ones you are about to intentionally create. Don't fool yourself: Fast Track Commercialization Projects create a rich set of problems. You are simply choosing which ones you will let happen and then fix by telling the DMAIC team what the gaps are likely to be on a proactive basis. You will "design" your problems.

- Training on critical parameter management. The team will need all the data possible to finish optimizing the product. The DMAIC team should be handed all CPM data, including the $Y = f(x)$ relationships that are underdefined. They will literally be finishing the critical parameter database.

- Robust design, tolerance design, and system integration and balancing competencies are a must in your post-launch DMAIC troubleshooting teams. These teams will have to complete tasks that were not fully done during commercialization. Some DMAIC training programs do not address these toolsets. You might need to transfer a few DFSS black belts into the post-launch DMAIC team to lead these tasks. Eventually, you will need to cross-train your DMAIC and DFSS teams so they share skills in these crucial areas.

Technology Development Capability to Support Fast Track Projects

If you are really good at developing stable, tunable, and robust technologies, you have the second foundational skill to justify rushing a select few commercialization projects. This is what Chapter 5 describes. If you are not good at developing these three things (stability, tunability, and robustness), rushing will get you in the end, in the post-launch environment where the DMAIC team gets the mess. If the project you want to put on a fast track depends on a new technology, you must pay your dues on it by not rushing the new technology through your R&TD process (IIDOV, as presented in Chapter 5). We don't see any way to rush both R&TD and commercialization; you must slow one down for the sake of the other. It is always better to get the technology done properly than take the risk of rushing the product in which it is embedded. The strategy we are trying to teach you is to take your time during portfolio renewal and R&TD; then you can rip through commercialization at a pretty good clip, depending on competitive pressures and market opportunities, to either rush or just hurry a product to market. If you are weak at portfolio management and technology development, you have to depend on your post-launch DMAIC skills alone.

Let's walk through the phases of the CDOV process (from Chapter 6, "Tactical Product Commercialization Process") and look at the options for tool-task groups that are a bare minimum of what will protect you as you rush a project. As a reminder, we focus on critical needs, requirements, and functions that are NUD—that is, new, unique, and difficult. They are the value-added elements that make the project worth doing.

Concept Phase Risk Profiles and Tool-Task Recommendations

Rushed projects have the following risk characteristics in the Concept phase:

- VOC needs content will be low because you have no time to dig down to get fresh VOC details. You have opinions, filtered input from marketing and sales professionals, trends you have picked up on, and secondary marketing data that is mainly historical in nature. *At a bare minimum, use KJ and QFD methods to structure and rank the data you have.*

- New technology critical parameter data will not be well linked to the current VOC data because it (VOC) is very limited. You might not have a good idea of performance-tuning ranges, so the tuning range for the new technology could be off course—hopefully not. If your new technologies are ranged properly, this risk will be low. Your technologies must be proven, with capability indexes and transfer functions that contain tuning capability, fully scalable to whatever target performance you ultimately need to launch the product for a given application. *Tunable and robust technologies make Fast Tracking Projects viable.*

- Competitive benchmarking will be low and likely anecdotal. This presents two areas of risk: missing technical requirement detail that would have come from deeper knowledge of what you are up against and improper ranking and prioritization of technical requirements because you are not sure what matters most. *At a bare minimum, define the key attributes and performance characteristics for the best competitive product you are up against.*

- Technical requirements will be ambiguous and changing as you move through the commercialization process. Targets and tolerances and their allocation down through the product's hierarchy will change because you will be discovering them as you go instead of using a stable set from the beginning, as you would in a nonrushed concept phase. *At a bare minimum, apply QFD to the requirements you have. This will be easier to change if what you have is at least well-documented.*

- Concept generation will be truncated to produce fewer alternatives than you would normally consider. Maybe you have only one concept and you bet the farm on it. If the technology enabling the single concept is stable, tunable, and robust, you have a better chance of seeing this one concept survive and make it through to an acceptable, but less-than-perfect launch. *At a bare minimum, try to structure two concepts from which you can hybridize a final product concept.*

- Concept evaluation and selection will be limited. Concept evaluation criteria will be anemic because of limited VOC data and the constraints placed on QFD and the translation of customer needs into stable technical requirements. Your concepts will suffer from added vulnerability to competitive threats and a lack of true superiority compared to alternative concepts. *At a bare minimum, use the NUD classification approach from your QFD work to stabilize the high-priority requirements.*

- Very limited system-level modeling, requirement budgeting and allocation, and early risk assessment of the system will be the norm. *At a bare minimum, use QFD NUD requirement flow-down traces to define NUD functions for the product.* With this in hand, conduct an abbreviated FMEA on the NUD functions to get a preliminary document of high-risk areas.

Marketing must do three tasks:

1. Verify the surrogate NUD customer needs by reviewing them with a few key customers for correction, revision, and validation.

2. Verify NUD technical requirements by reviewing them with a few key customers for correction, revision, and validation.

3. Verify final product concept by reviewing it with a few key customers for correction, revision, and validation.

All this work is done to quickly check whether there are any major misses. This is just the bare minimum, to be sure, on a "check back with the customer" basis. Normally, you would take more time and really dig in with a broader set of diverse customers to gather the data from them proactively. Here, internal opinion and past experience generate surrogate VOC data. You can justifiably claim you are customer driven with this approach; it's just that the cart is in front of the horse!

Summary: What Must Be Done at a Bare Minimum in the Concept Phase?

- Document fresh needs data from your marketing and sales experts that gives you their best field experience and judgment on customer needs. You are basically using them as a surrogate for the real VOC. Apply KJ to structure, rank, and prioritize this surrogate customer needs data. Check back with a few key customers to be sure you are reasonably representing their latest needs.

- Document the characteristics of the best product in the market that you know from your internal opinion. What will present the biggest threat to your new product? This is surrogate benchmarking with no actual tear-downs or in-depth analysis of various competitive products.

- Use KJ and QFD to process your limited data sets to structure, rank, and prioritize the customer needs for translation into a documented set of product technical requirements. Check back with the same few key customers to be sure the technical requirements are basically aligned with the NUD customer needs.

- Document at least two alternative concepts to compare against the one best product you are going up against. Use the NUD requirements from the House of Quality to help establish concept evaluation and selection criteria.

- Select the best of the two concepts and refine one concept to establish the most competitive concept possible to go up against your competitive benchmark. If you can, blend synergistic attributes from both concepts to enable one superior concept. Check back with the few key customers again to be sure they are aligned with where this product is headed.

- Conduct an FMEA on the NUD functions within your selected product concept to characterize risk.

Key Tools, Methods, and Best Practices to Consider as You Control Risk

- KJ method for structuring, ranking, and prioritizing customer needs
- QFD and the House of Quality
- NUD screening of customer needs and technical requirements
- Pugh concept evaluation and selection process
- Design FMEA

Design Phase Risk Profiles and Tool-Task Recommendations

Rushed projects have the following risk characteristics in the Design phase:

- Stability of functional performance inside the subsystems and subassemblies absolutely must be proven and documented.

This is especially true of any influences new technologies have on leveraged designs. New materials are particularly important if they are used in existing hardware form and fit structures. Here the documentation that proves stability for each critical sublevel function is the SPC chart known as an Individuals and Moving Range Chart (I-MR chart).

- The next level of risk is in the area of functional performance capability under nominal or best-case conditions. Here you must insist on documenting Cp and Cpk values for critical sublevel functions. The focus is on nonstressed performance— what we call nominal conditions where there are only random sources of noninduced variation active in the data. Here, capability studies are required for all new, unique, and difficult functions that are the basis of customer satisfaction in the sublevel designs.

- The final required documentation for this phase is tunability. What one critical adjustment parameter (CAP) is proven to have a statistically significant effect on shifting the mean of each critical functional response in each sublevel design? If we were not rushing, we would take the time to define numerous significant CAPs, not just one dominant CAP. You are *not* building a comprehensive Y as a function of multiple Xs here— just one simple regression model for which you spend only enough time to nail down one critical X for tuning the mean performance onto the desired target.

Summary: What Must Be Done at a Bare Minimum in the Design Phase?

- You must document baseline performance stability and capability at the subsystem and subassembly levels. If the designs do not possess stability and acceptable capability under non-stressful conditions, you do not have the building blocks needed to integrate a product in the Optimize phase. The design's critical Ys should also be tunable to hit the targets defined in the product requirements document. If you cannot scale functional performance at the sublevels, you cannot adjust and balance functional performance at the system level. Without this, you simply do not have a viable product by any standard of measure.

Key Tools, Methods, and Best Practices to Consider as You Control Risk

- Statistical process control for stability of performance (I-MR charts)

- Regression analysis for documenting key mean-shifting parameters (CAPs)

- Capability studies under nominal conditions

- Design FMEA to update risk for the NUD functions that made the design worth doing from a customer's perspective

Optimize Phase Risk Profiles and Tool-Task Recommendations

Rushed projects have the following risk characteristics in the Optimize phase:

- Here you do just one robustness optimization screening experiment per sublevel design *if* it possesses an NUD function that is critical to customer satisfaction for this new product. We prefer to iterate through a sequence of robustness experiments for all our major subsystems and subassemblies, but a quick screen for critical Xs that have a stabilizing influence on the standard deviation of each critical Y under nonrandom, induced stress conditions is a must. If you don't do this at a minimum, you definitely will be slowed during system integration, and you just cannot afford to let that happen.

- When we have the NUD subfunctions somewhat robust to stressful sources of variation, you can take the risk of rapid system integration. Here you run one nominal, nonstressed performance test and document Cp and Cpk for the critical Ys at the sublevel *and* system levels. Then you run just one stress screen using a designed experiment to find out where the system integration sensitivities reside. You tune the sublevel design parameters to try to balance performance under stressed conditions, thus minimizing k shifting in the mean performance of the sublevel Ys and, consequently, stabilizing the variability of the system-level Ys that are critical to the customer and your business case.

Summary: What Must Be Done at a Bare Minimum in the Optimize Phase?

- Conduct limited stress tests on the customer-critical subsystems and subassemblies to flush out sensitivities that can be rapidly identified and fixed. The best thing you can do is give the system as robust subsystems and subassemblies as you can afford—in our opinion, at least one round of robustness screening. This helps make integration go faster and smoother. The less work you do here, the more trouble you will have during system integration and final product verification. Do as much robustness optimization as you can here. Let the DMAIC teams focus on the final tuning and optimization of mean performance; they are better at that kind of work. Your goal is to tame the standard deviation during intentionally induced stress evaluations.

- Conduct a limited system-integration stress test to identify sensitivities that can be rapidly identified and balanced. Here you make a first cut at preventing escaping integration problems. You won't get them all, but you'll get the really bad ones out of the way, depending upon how much stress you choose to induce.

- Document failure modes for the latest data so the DMAIC team can see emerging risk trends and tasks you did not complete. They have to structure their project plan to fill your gaps.

- Conduct process FMEA to identify risk in the relationships between design functions and process functions that make the parts and materials that will affect the design performance, fit and finish Cp & Cpk values.

Key Tools, Methods, and Best Practices to Consider as You Control Risk

- Robust design of experiments

- Post-robustness SPC charts

- Capability studies before and after robustness stress evaluations (Cp and Cpk indexes)

- Design and process FMEAs to link process risk to design performance risk

Verify Phase Risk Profiles and Tool-Task Recommendations

Rushed projects have the following risk characteristics in the Verify phase:

- NUD tolerances and nominal set points down through the integrated system must be finalized. Here you use tolerance analysis aided by Monte Carlo simulations and designed experiments to balance the integrated system, with a constrained focus on the new, unique, or difficult functions that matter most to customers. Careful identification of just those set points at the sublevel design arena and system output should be targeted for focused tolerance optimization. Let the rest of the set points be specified for what the supply chain can reasonably hold at the most reasonable price. Let the post-launch team clean up the final specifications that have to be balanced for cost vs. performance.

- Final reliability evaluations can be run at this time to screen for limitations and focused corrective action. We know that postponing a lot of system-level reliability work until this final phase is scary, but we think it is a better trade-off to spend your limited time on robustness development than just depress yourselves with premature reliability assessments that do nothing to improve reliability. They chew up precious prototype hardware evaluation time and human resources. You already know you're intentionally developing a less-than-ideal product for rapid launch; let the post-launch team apply aggressive reliability growth tasks in harmony with cost vs. performance balancing. If their DMAIC skill are strong, this work will be reasonably quick and straightforward for them vs. robustness development tasks that are far more specialized to DFSS-enhanced commercialization teams.

Summary: What Must Be Done at a Bare Minimum in the Verify Phase?

- Conduct tolerance-balancing analysis and designed experiments to finalize specifications for the NUD functions that matter the most to customers.

- Conduct capability studies at the system level on the product to identify current, integrated performance. The DMAIC team will need this data to begin improvement projects on the weak areas across the system.

- Conduct reliability assessments. Document failure modes for the latest data so the DMAIC team can see emerging risk trends from the tasks you did not complete. The team has to structure its project plan to fill your gaps.

Key Tools, Methods, and Best Practices to Consider as You Control Risk

- Analytical tolerance design (Monte Carlo simulations for tolerance balancing)

- DOE for integration sensitivity analysis

- Reliability assessment

- HALT and HAST testing (only if you have time)

- Capability studies

- Design FMEA

- Process FMEA

Macrosummary of Fast Track Projects

Fast Track Projects can lead to helpful additions to sustainable growth. They alone cannot produce sustainability; only a balanced portfolio of new products can accomplish that goal. The non-negotiable, critical few requirements for a Fast Track Project include the following:

Conceptual design phase:

- Clear and stable requirements that are derived, as much as possible, from real customer data. In this case, it will be mainly "check back with the customer" oriented. You are using needs that are truly new, unique, and difficult (NUD) to drive criticality vs. easy, common, and old requirements.

- A superior product concept that was hybridized out of at least two alternatives that were compared against one best-in-class benchmark. The concept evaluation and selection criteria must be based upon the NUD requirements that are traceable back to real customer preferences and needs.

- A risk assessment based upon the potential failure modes of the product, the market, and the project-management tasks that are being planned to complete it.

Nominal design development phase:

- Baseline stability and tunability of all critical (NUD) sublevel design elements

- Cp and Cpk data for all critical functional responses from the sublevel designs under nominal, nonstressed conditions

- A risk assessment based upon the potential failure modes of the sublevel designs, the market, and the project-management tasks that are being planned to complete the next phase

Optimization and integration phase:

- Sublevel design robustness screening to identify sensitivities and parameters that diminish their affect on the standard deviation of the NUD critical functional responses.

- System integration tests under both nominal and stressed conditions, prior to system-level reliability-assessment tests. System-level sensitivities are diminished and balanced across the NUD critical functional responses.

Product and production system verification phase:

- Sublevel and system-level NUD tolerances balanced for cost and performance

- Production tolerances and nominal set points capable of supporting the NUD functional responses within the integrated product

Boiling this all down, you can see that you truncate your tasks to a bare minimum that is totally documented by critical parameter management for just the new, unique, and difficult requirements and functions of your Fast Track Product. The post-launch team will deal with the interaction of the NUD functions and the easy, common, and old (called ECO) functions. If you get the dominant sensitivities under control for the NUD functions, you minimize the damage of poor performance across the items that really make the product value-adding to the customer. The DMAIC team will have to quickly clean up the remaining sensitivities and reliability issues.

Pick—or, better yet, design—your problems wisely and proactively; be sure your DMAIC team is well aware of its role in this kind of project; and don't hope for the best—*plan* for it! Proceed with caution: This is not the way you should be commercializing products as a rule. Do this only as the exception, when you can prove the risk is worth taking from good portfolio management discipline, as outlined in Chapter 4.

8

OPERATIONAL POST-LAUNCH ENGINEERING SUPPORT PROCESSES

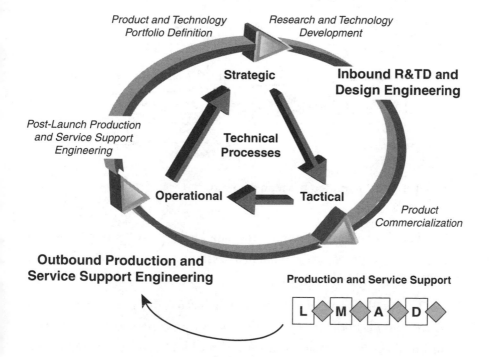

Post-Launch Product, Production System, and Service Support Engineering Six Sigma in the Operational Technical Process

We are at the end of our integrated process flow at this point (see Figure 8.1). We have used our strategic processes, product portfolio renewal, and research and technology development to prepare projects with discipline and rigor for either normal, lean commercialization or Fast Track Project commercialization. Technical workflow design for launching, managing, adapting, and discontinuing commercialized projects is now our focus.

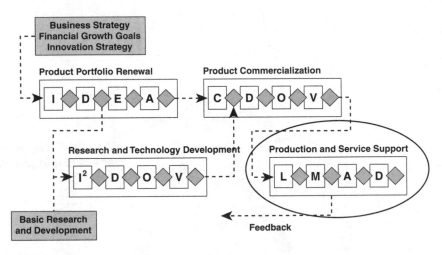

FIGURE 8.1   Integrated process diagram.

The flow of critical parameter data back upstream to the tactical and strategic processes and teams is essential for this system to work. Without it, we don't learn to prevent problems in the future.

Let's look closely at what Six Sigma has to offer in the post-launch environment, both from a steady state process perspective (LMAD) and in the special, react-to-a-serious-problem (DMAIC) context we are so familiar with from an historical Six Sigma perspective.

Production and Service Support

The LMAD process is for preventing problems, the DMAIC process is for reacting to them when they have become so bad that we can detect their effect on our financial, cycle-time, or quality metrics. Our discussion mainly focuses on measuring things in the post-launch environment that help us curb problems before they force our hand and require a formal DMAIC project. If you are not seeing a decline in DMAIC projects as time goes on, you are not doing Six Sigma and Lean properly. The things you must measure to snap out of this reactive paradigm are called leading indicators. We are all pretty good at defining and measuring lagging indicators—they are easy to get at, but are lousy at preventing problems.

While conducting the work required for meeting technical launch objectives, it is easy to get caught up in a vast sea of unexpected variation. Plans are made, and then, with the turn of the hands on the clock, everything seems to be changing. Technical support teams can go "off plan" fast. In Six Sigma terms, this means assignable causes of variation (noise) cause the technical processes to go out of control. It is easy to be tempted to react to whatever comes along. Significant levels of planning seem to be a waste of time because things come along and cause us to change the plan—so why bother planning? The plan is often just, "Be good at reacting." This is why DMAIC gets "leaned on" improperly in some Six Sigma–oriented companies. They use DMAIC strategy as all that Six Sigma really is and has to offer. DMAIC methodology is very valuable in the post-launch environment; it is a great way to find and fix problems methodically. As you learned in Chapter 7, "Fast Track Commercialization," you can use it tactically to help rush a project. In the normal environment of nonrushed, lean product-commercialization projects, you can borrow some tools and methods from the DMAIC toolset. These often are used as part of the control plan for launching, managing, and adapting a product. But they and the DMAIC steps are not *the* control plan across the post-launch lifecycle.

Change in the operational environment of post-launch technical support is normal. If you are surprised by it, you probably haven't been doing product launches for very long. To be good at sustaining growth, you must anticipate changes by learning to measure things that can help your team adjust the flow and nature of your technical support tasks. You need to be able to design a plan that is adaptive by design, so you really never have to go "off plan." This is where critical parameter management and the database developed during R&TD and commercialization are so helpful. Without them, you are flying blind into an environment where change will be present almost all the time. We illustrate this by examining the cycling between the Manage and Adapt phases as we go along the LMAD process.

What does it mean to anticipate change? It means to work diligently to identify and measure leading indicators of market, competitive, and customer behavioral dynamics, as well as supply chain, production, assembly, and service process dynamics that prepare your team to make preplanned changes in your tasks and tools—*but not your plan itself!* We can frame this proactive measurement approach by using the injunction, "Measure impending failures."

A simple analogy helps illustrate this point. Heart attacks are a leading cause of death in the adult population (a "failure" we can count). Obesity is a measurable attribute that is a proven link to causing a heart attack (a "quality attribute" we can measure). Food intake and amount of exercise are measurable functions (not quality attributes) that are proven links to obesity (gross fundamental measures of causes of obesity). The specific kind of food one eats, how much exercise one gets, and the time and type of exercise conducted are basic, fundamental measures of impending failure. They are measurable, controllable, adjustable factors that can prevent or slow the rate of the onset of obesity and ultimately control the risk of a heart attack. If we plan our diet and exercise regime carefully, we can manage and adaptively control our lifestyle to lower our risk of heart attacks—even if we are genetically disposed to heart problems. Of course, this requires discipline to measure. You have to measure the fundamentals that are directly causing a functional, traceable, technical dynamic that eventually leads to "failure." Obesity and heart attacks are lagging indicators. We can only react to them after they are detected.

Don't wait for lagging indicators to scare you into action after they have already developed to the point that you can finally detect them. Measuring the fundamentals of human behavior, decision-making patterns, purchasing trends, manufacturing variations and mean shifts, supply chain dynamics, and product use behavior will set you on a course of action that lets you prevent and reduce technical process failures. To get good at this approach, you have to become a student of post-launch "noises."

A "noise" is any source of variation. In technical processes, they are normally caused by the laws of physics doing things we wish they did not. They are not the relatively small, random variation that is present in all forms of technical objects, designs, systems, and processes. They are assignable causes of variation that the methods of robust design were created to deal with. They often precede an outright failure in a process. They are changes that affect the mechanics, the fundamental cause-and-effect relationships inside technical functions and support processes. Processes contain physical steps that we or machines do to produce a targeted result. Noises alter the targeted results we are attempting to achieve on a sustainable basis. They disrupt the fundamental functions that control measured results. In this chapter, we explore how to apply Six Sigma discipline proactively in the presence of such provocative noise during product launch, management, adaptation, and final discontinuance.

In the operational technical process environment, we have identified a generic process segmented into four distinct phases. The LMAD process for controlling the lifecycle of the portfolio of launched products:

1. **Launch** the product through its introductory, scale-up period into the steady-state production, use, and service environment.

2. **Manage** the product in the steady-state technical support processes where stability and capability are our major concerns.

3. **Adapt** the technical support tasks and tools as "noises" require change. Here the major concern is tunability and robustness.

4. **Discontinue** the product and associated production and service processes with discipline to sustain quality, ongoing brand loyalty, and margins, and minimize costs.

New here is our focus on not waiting until a lagging indicator of performance goads the team into chartering a special DMAIC problem-solving project. We use the LMAD phases to establish a proactive system of critical technical parameters, so we limit the number of problems by design. LMAD phases, enabled by preplanned Lean, Six Sigma technical support tasks and tools to prevent problems, replace the need for emergency DMAIC steps to react to problems. Think of it as a health club, not an emergency room. We prefer to avoid heart attacks.

The notion of a gate review that we have grown accustomed to from strategic and tactical technical processes over the previous four chapters must change now that we are in the post-launch environment. Operational technical processes are not like portfolio renewal, R&TD, or product commercialization processes. As in steady-state manufacturing, no formal gates exist in this environment. Phases occur in time as a function of market and technical dynamics. The Launch phase usually occurs just once—unless, of course, you rush through a commercialization project without following the advice given in Chapter 7. Then you might see relaunches, as I have—never a happy experience, by any measure. The phases that recur are Manage and Adapt. Depending on the lifecycle of the product, many cycles of Manage and Adapt might occur. There should be pre-planned, periodic reviews of progress against the technical support requirements as they link to sales, margin growth, and cost targets. These should be thought of as key milestones in the continuum of executing the technical support plan, not as gate reviews. Data and results can be gathered and summarized on an hourly, daily, weekly, monthly, quarterly, and yearly basis. It is essential that you balance proactive measures of performance with reactive measures of lagging indicators of performance. Mostly you want to focus on leading indicators of performance if you want to stay on plan and under control. You have to be realistic and recognize that it is almost impossible not to measure some lagging indicators; you just want to keep them to a minimum. This is part of what it means to be "lean" in terms of critical parameter management.

The work of "outbound" technical support teams is conducted in post-launch product line management. A distinct set of measurable results comes from this work. The technical tools, tasks, and deliverables are all defined by the requirements for launching, managing, adapting, and discontinuing a product (see Figure 8.2).

FIGURE 8.2   Phase-milestone flow for the LMAD process.

Most of the tools, methods, and best practices of "outbound" technical support are the same as those used in the later phases of commercialization. We focus on the ones that add value to staying on plan. The difference is that the technical support professionals in the operational environment are using these tools to *refine* specifications, *adjust* key parameters that affect mean performance, and *analyze* data streams for adaptive purposes when cost and performance goals are in jeopardy of not being met.

The "lean" way to gather data and make decisions in the post-launch environment is to use leading indicators of performance against plan. Measuring lagging indicators must be done in some cases, but let's be honest: They only let you know when you have to react to a problem they have detected. Fire detectors work. The house burns while the fire department gets to the scene to contain the damage. Smoke detectors are better. They might warn early enough to allow the smoldering source of fire to be extinguished before there is significant damage. Best is a planned day-to-day way of behaving that prevents the fuel, air, and ignition sources from getting into the right "dynamic" combinations to start a fire in the first place. Lean technical support teams focus on their forms of fuel, air, and ignition sources so that the fundamentals, the dynamic process parameters, are measured and under adaptive control, preventing smoke and fire as much as possible.

Early warning data types and structures must be invented and used to enable a proactive approach to preventing technical support problems. Much of this work is done during commercialization and then applied during LMAD phases. This requires innovation of measurement systems and data structures by the commercialization teams and then their planned and controlled transfer to the post-launch teams. The best teams measure things their competitors do not even know exist as measurable forms of data. Some competitors might be aware of these metrics but find them difficult to measure. Let's hope they are lazy and use DMAIC as their plan. We must face this struggle to measure impending failures and prevail. Let's change from focusing on detecting defects and shift to vigilance in measuring leading indicators of impending failure. If we don't, we won't be able to sustain growth—just like everyone else.

Hard vs. Easy Data Sets

Effective control of the LMAD phases depends on what we choose to measure and how often. Many business processes have been designed and fitted with measurement systems that are loaded with things that happen to be easy and convenient to measure. Often data that is easy to get is least effective in telling you anything fundamental about what is really going on in your technical post-launch environments. Easy-to-measure data is almost always a lagging indicator of what has just happened to you. Just how lagging a measure can be, is a function of how often you choose to take data and when you get around to analyzing it. It is common to hear teams state that they have tons of data but lack the time and resources to analyze it. This is where "lean" methods can help. Technical critical parameter management takes the hard fork in the road when it comes to the type of data and the number and frequency of samples. We need to gather the right data and just enough of it to make a statistically sound decision. If you measure attribute data, you must gather large amounts of it to make a sound statistical decision. If you measure continuous variable data, typically associated with a fundamental, technical dynamic, then relatively small samples of data are needed to make a statistically sound decision.

A good example of a poor measure of supply chain performance influence on revenue is the number of supplier quality assurance professionals sent out to interact with vendors. Sheer numbers of people is a gross measure that is easy to quantify but is not at all capable of providing critical parameter relationship data in the context of how the actual production processes are working. True process dynamics data is just plain hard to gather, document, and communicate—but it is a set of parameters that is fundamental to quality. If you can get it, it will show a fundamental cause-and-effect relationship.

To stimulate innovation around data integrity and utility, you need to ask a few simple questions:

- Are you measuring continuous variables in your technical processes that are fundamental to the dynamics taking place within these processes?

- Are you measuring variables that your competitors are not aware of or are ignoring because of difficulty in gathering such data?

Critical parameter data is the result of taking selected, focused measures while doing the work of making materials, parts, subassemblies, and so on. This lean set of data is produced and gathered by conducting select tasks using specific tools, methods, and best practices. A common set of tools, methods, and best practices is used repeatedly during the LMAD phases. These are discussed next and help enable the tasks of technical support teams.

The Tools, Methods, and Best Practices That Enable the LMAD Tasks

Technical support process definition:

- Technical support process requirements development
  - Customer and partner interviewing and requirements data gathering
  - Requirements structuring, ranking, and prioritization (KJ analysis)
    - Quality function deployment (defining detailed metrics)
- Technical support process mapping
  - Technical functions, inputs, outputs, and constraints
  - Technical process noise mapping
  - Technical process failure modes and effects analysis

- Concept generation for technical support professionals
- Pugh concept evaluation and selection process
  - Technical support process concept innovation
- Technical support process critical parameter management
  - Product and production system critical parameter identification and metrics
  - Service process critical parameter identification and metrics
  - Supply chain critical parameter identification and metrics
  - Customer relationship and support critical parameter identification and metrics

Technical support process risk management:
- Product management scorecard design and applications
  - Risk analysis, management, and decision making
- Technical support process failure modes and effects analysis
  - Risk analysis, management, and decision making
- Technical support SWOT analysis

Technical process models, data analysis, and controls:
- Product line management control planning
- Project-management methods
  - Cycle-time Monte Carlo simulation
  - Critical path task FMEA
- Technical process cost modeling
- Product and service cost modeling
- Service models and parts forecasting
- Market perceived quality profiles (MPQP)

- Service customer value-chain mapping

- Design of surveys and questionnaires (for service)

- Post-launch technical support data structures, analysis, and management

  - Descriptive and inferential statistics for technical support data

    - Graphical data mining

    - Multivari studies

    - Hypothesis testing, confidence intervals, t-tests, data sample sizing

    - Regression and model building

  - Capability studies for technical data

  - Statistical process control for technical data

  - Design of experiments for technical parameters

    - Fractional factorial designs

- Adaptive cost, reliability, and service/maintenance forecasting methods

  - Monte Carlo simulation

- Data feedback structures for advanced product and production planning

- Product, production system, and service discontinuance planning

With the tools, methods, and best practices categorically defined, let's look at the phases of the LMAD process and explore how we can conduct technical support tasks with their help.

The LAUNCH Phase

The technical support teams have generated an integrated plan to see the launch through its cycle. The LAUNCH phase is concluded when the product, the production system, and the associated services and support initiatives are fully established, available, and in a steady-state mode for the MANAGE phase. The launch plan has a well-defined set of measurable control variables, functions, and results that are fundamental to the behavioral and technical dynamics that characterize the unique issues associated with launching a new product, production system, and service network. The consistent application of a designed launch-control plan has built-in robustness features to prevent excessive sensitivity to assignable, nonrandom sources of variation in the launch environment. Launches are quite dynamic and are frequently affected by scale-up noises.

The *requirements* for the LAUNCH phase milestone include the following:

- Product, service, and production system support resources are fully operational and stabilized.

- Standard operating procedures are in place and proven, with data, to be working properly when compared to the detailed post-launch technical support requirements.

The *deliverables* for the LAUNCH milestones include these:

- Production and service capability studies

  - Actual production and service data vs. service forecast

- Initial production and service trend assessment

- Production and service capability growth rate vs. plan

- Vendor identification and qualification metrics

- Vendor experience and behavioral analysis

- Vendor satisfaction assessment

- Vendor failure modes and effects analysis

- Competitive assessment

- Vendor effectiveness evaluation

- Technical support critical parameters update

  - SPC charts

- Business case fulfillment assessment

- MANAGE phase control plan documented and ready for use

The *tasks* within the LAUNCH phase include these:

1. Gather critical product, production system, and service data.

2. Analyze critical parameter performance data.

3. Generate statistical process control charts and capability indexes for key response metrics using the data sets.

4. Refine reliability and capability forecast models.

5. Generate LAUNCH phase risk assessment.

   - Conduct ongoing FMEAs within the LAUNCH phase technical processes.

6. Assess business case against current performance.

7. Update MANAGE phase control plan.

The MANAGE Phase

When the product, production system, and service process has gone through its planned launch cycle, the technical team follows a steady-state control plan to manage these three entities. A steady-state technical support plan consists of key tasks that are enabled by specific tools, methods, and best practices to sustain the product, production system, and service process as it enters a mature, "scaled-up" environment. Technical support process critical parameters are measured and evaluated to determine when the ADAPT phase is required to sustain their respective cost, quality, and performance goals.

This is the *requirement* for the MANAGE phase milestone:

- Technical support processes, their detailed functions, and measurement systems are capable of detecting assignable causes of variation that indicate when adaptive actions are required to stay on plan.

The *deliverables* for the MANAGE milestone include the following:

- Product, production system, and service process capability studies
  - Data vs. forecast
  - Steady-state capability trend assessment
  - Capability growth rate vs. plan
- Vendor capability metrics
- Vendor experience and behavioral analysis
- Customer satisfaction assessment

- Technical support process failure modes and effects analysis

- Competitive assessment

- Vendor effectiveness evaluation

- Technical support critical parameters update

  - SPC charts

- Business case fulfillment assessment

- ADAPT phase plan documented and ready for use

The *tasks* within the MANAGE phase include these:

1. Gather critical technical support data.

2. Analyze critical parameter performance data.

3. Generate statistical process control charts and capability indexes for key metrics using the data sets.

4. Refine reliability and capability forecast models.

5. Generate Manage phase risk assessment.

   - Conduct ongoing FMEAs within the Manage phase technical support processes.

6. Assess business case against current performance.

7. Refine ADAPT phase control plan.

The ADAPT Phase

A technical support plan is in place that has a wide variety of contingency plans, enabled by specific sets of tools, methods, and best practices that control the critical technical support parameters available to counter assignable causes of variation that disrupt the steady-state plan for supporting sales, margin growth, and cost targets. The ADAPT phase can be exited by bringing the technical support metrics back to a steady state of control after the effects of variation have been countered. If requirements have changed during the MANAGE phase, the adapted outputs must meet them.

Sources of assignable cause variation (nonrandom noises) signal the technical support teams to adapt or make adjustments to critical adjustment parameters that are known to be capable of countering these effects. These variations are typically changes in product use, production system, or service process dynamics or competitive market dynamics that threaten your results. The goal is not to change your plan, but to adaptively use it to stay in control of parameters that help sustain growth.

The *requirements* for the ADAPT phase milestone include the following:

- Identify critical adjustment parameters that can return the technical support processes to a state of statistical control.

- Identify the leading indicators (assignable causes of variation) that signal the technical support teams to adjust their critical adjustment parameters.

The *deliverables* for the ADAPT milestone include these:

- Technical support critical parameters update

  - SPC charts

  - Critical adjustment parameters

- Updated technical support process noise maps

- Updated technical support process FMEAs

  - DISCONTINUE phase plan documented and ready for use

The *tasks* within the ADAPT phase include these:

1. Apply critical adjustment parameters to adjust critical technical support functions and get their results back on target.

2. Conduct designed experiments as necessary to improve effectiveness of critical adjustment parameters for current operating conditions.

3. Generate statistical process control charts and capability indexes after adjustments.

4. Refine cost, capability, reliability, and performance models.

5. Refine supply chain management plans.

6. Generate ADAPT phase risk assessment.

7. Update competitive and supply chain situation analysis.

8. Conduct SWOT analysis.

9. Refine FMEAs and noise maps based upon results from critical parameter adjustments.

  • Conduct market perceived quality profile and gap analysis.

10. Assess business case against current performance.

11. Refine ADAPT phase control plan.

12. Refine DISCONTINUE phase control plan.

The DISCONTIUE Phase

As the product and its production and service processes near their planned or forced end of life, a preplanned set of deliverables, tasks, and tools will increase the likelihood of an efficient and cost-effective transition to new products, production systems, and services as your product portfolio matures and is renewed.

The *requirements* for the DISCONTINUE phase milestone include the following:

- Define discontinuance criteria.

- Define market, sales, service, supply chain, and production system conditions that fit discontinuance criteria.

- Conduct product, service, supply chain, and production system discontinuance according to the discontinuance plan.

- Provide discontinuance data to the product portfolio renewal team.

The *deliverables* for the DISCONTINUE milestone include these:

- Documented discontinuance criteria

- Critical technical support parameter data

  - SPC charts

  - Capability studies

  - Trend analysis

- Discontinue phase risk assessment

  - Technical support process FMEAs

  - Technical support noise maps

  - Competitive assessments

- Business case assessed against current performance

- Recommendations for sustaining the supply chain, service process, and production system through next-generation product

  - Voice of the marketing team data

  - Voice of the customer data

- Voice of the sales team

- Voice of the supply chain

- Voice of the service team

- Voice of the production system team

- Market perceived quality profile and gaps

- Final SWOT and Porter's Five Forces analysis

- Lessons learned document

The *tasks* within the DISCONTINUE phase include these:

1. Apply critical adjustment parameters to adjust critical technical support functions and their results, to control the discontinuance of the product and services to protect the brand and maximize the business case.

2. Conduct designed experiments as necessary to improve the effectiveness of critical adjustment parameters for discontinuance conditions.

3. Refine cost models for discontinuance.

4. Refine supply chain management plans for discontinuance.

5. Generate discontinuance service and support forecast models.

6. Generate and provide technical support data to the product portfolio renewal team.

   - Conduct SWOT and Porter's Five Forces analysis.

   - Conduct MPQP and gap analysis.

   - Conduct competitive assessment.

   - Conduct voice of X data gathering.

   - Generate lessons learned document.

7. Conduct final business case assessment against goals.

The LMAD process is now complete. We have illustrated how Six Sigma can be used to help control, adapt, and refine what we do and when we do it in these phases and the key milestones of each phase. This work can be easily designed to be lean, customer focused, and value-adding. It is worth noting again that the Manage and Adapt phases can repeat numerous times before discontinuance is conducted. How long a product and its services can generate revenue and margin in fulfillment of the business case is a variable that can be forecast and assessed with the help of the Six Sigma tools, methods, and best practices.

When unexpected problems occur and cause a real crisis, it is the right time to structure and launch a DMAIC project to fix the problem. If the LMAD process or a subprocess within it is no longer capable of supporting the growth goals of the business, it can be overhauled or redesigned using the DMADV process.

As in all the other processes, you can use the tool, task, and, now, milestone scorecards to quantify the performance of post-launch technical support teams. Refer back to Chapter 2, "Scorecards for Risk Management in Technical Processes," for their general use instructions.

TABLE 8.1 Milestone Deliverable Review Scorecard

1	2	3	4	5	6
Milestone Deliverable	Grand Avg. Tool Score	% Task Completion	Results vs. Requirement	Risk Color Code(R-Y-G)	Milestone Requirement

TABLE 8.2 Task Scorecard

1	2	3	4	5	6
Task	Avg. Tool Score	% Task Completion	Task Results vs. Required Deliverable	Risk Color Code(R-Y-G)	Milestone Deliverable

TABLE 8.3 Tool Scorecard

1	2	3	4	5	6
Tool	Quality of Tool Use	Data Integrity	Results vs. Required Task Deliverable	Specific Tool Score	Task & Deliverable

9

FUTURE TRENDS IN SIX SIGMA AND TECHNICAL PROCESSES

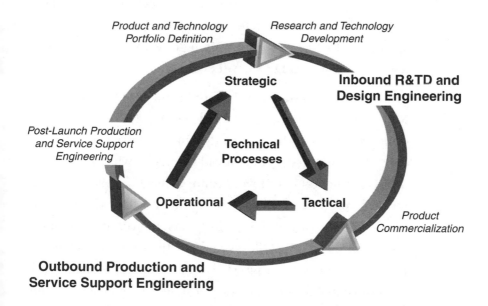

Trends in Lean and Six Sigma for Technical Processes

A few trends likely will occur as Lean and Six Sigma have an impact on technical processes for sustaining growth:

- **Investment in Training:** More money will be invested in specific value-adding technical functions (clusters of tool-enabled tasks) that produce deliverables that underwrite the success of growth initiatives. Training will focus not on one tool, but rather on linked tool-task groupings that are very tightly integrated with the technical phase-gate processes in which they are embedded. Functional excellence starts at the college or university level. It matures in the older form of continuing education, most commonly seen as 40 or more required hours a year in some form of technical training. This paradigm will be recognized for its major shortcoming: poor linkage to tasks, deliverables, and phase-gate process governance and risk management. More programs of training directly linked to project application, such as DFSS, will evolve.

- **Measuring Fundamental Indicators of Performance:** More fundamental technical variables ($Y = f(X)$) that are truly critical to customer behavioral dynamics will be measured and controlled. A better definition will emerge of what underwrites a real cause-and-effect relationship. Critical technical parameters that prevent problems will be identified and measured. Variables that signal impending failure will be measured instead of measuring failures and reacting to them. Technical teams will stop measuring what is easy and convenient if it is not fundamental to true cause-and-effect relationships within and across integrated systems of technical variables.

- **Increased Investment in In-bound Marketing:** Enterprises will increase the number of marketing professionals, in balanced proportion to engineering professionals, so they can do the right tasks fully and completely. This will reflect the designed balance between your marketing and technical innovation strategies. Plan on conducting more collaborative frontend work where inbound marketing professions work closer than ever with technologists and design engineers to drive customer needs deep into the structure of technical requirements. Also plan on seeing this same increase in collaboration between outbound marketing and sales professionals as more fundamental and preventative data sets and metrics are driven into the post-launch product line–management operations.

- **Less Multi-tasking:** Technical work flows will be differentiated into strategic, tactical, and operational "buckets": product portfolio, R&TD, product design, and production/service support engineering organizations will work in harmony and center on critical parameter data development, flow, and cross-boundary transfer. Jack-of-all-trades technical professionals are a major factor in incomplete technical tasks and poor use of technical tools. Just study the characteristics of the technical workforce at Toyota, and you will see many companies migrating to that set-based, systems approach to product development. Too little work gets done by the wrong people. Great R&TD professionals don't always make great product-design or production-support professionals. You can also expect to see far less ladder climbing where superb technical professionals struggle to get into a career path to upper management. Incentives to stay technical and to stay in the area of expertise will increase. When properly rewarded and given appealing work conditions and management support, many bright, innovative technical professionals will stay put and get really good at their craft.

- **Focused Core Competetence Centers:** Technical centers of excellence will be formed to promote continuous improvement and the standardization and "right-sizing" of technical tools, tasks, and deliverables. Workflow standardization will replace tool standardization. Dominating how technical teams design their work will be not what the tools can offer, but rather what tool-task groups can deliver relative to gate requirements.

- **Problem Prevention vs. Problem Solving:** A shift will take place from DMAIC Six Sigma for problem solving and cost control to a phase-gate approach to problem prevention and investment in properly designed technical workflows to enable growth projects. A reversal of back-end-loaded, unbalanced investment in the reactionary environment of post-launch problem-solving teams will occur. Companies will spend more up front to prevent problems. DMAIC emphasis and use will diminish.

- **Collaborative Innovation:** Far more collaboration will take place between marketing and technical teams and professionals across strategic, tactical, and operational environments. "Not my job" attitudes will be replaced with better cross-functional teamwork. This is particularly true in all areas where customer needs, complaints, and sensitivities require translation into technical requirements to avoid future problems by proactive knowledge sharing.

- **Strategic Platforming:** Technical and marketing skills must migrate to platform and modular design thinking to help design and meter the flow of product and service offerings in a balanced portfolio-deployment context. Not every product can be loaded with maximum features and functions if we expect to be on time in product launches. An elegant, designed flow of product families and preplanned line extensions makes it much easier to align limited corporate resources to rapidly evolving market and competitive dynamics. It also reduces the intensity of risk if we happen to miss one of our products because we are not "betting the farm" on a routine basis. A very few well-defined projects can be rushed, by design, to market. Increasingly fewer businesses will use an "all hands on deck—rush!" approach to commercializing all their products.

- **Realistic Design of Cycle-time:** Cycle-time hubris will decline. I want you to look up the word *hubris* so that its full impact reaches your cerebral cortex. If you have it, relative to your people's workload and cycle-time capabilities, you must root it out of your consciousness and pay closer attention to what it really takes to get work done right the first time. Technical executives must enrich and expand their current level of understanding of how things get done to a more fundamental appreciation of tool, task, and deliverable workflows. It takes time to do things right; executives who want to sustain growth will have to get in touch with the "design" of cycle-time. When executives get aligned with the proper structuring (right-sizing) of technical work, the efficiency and performance of technical teams will become far more predictable. Stress levels need to be lowered by facing the fact that technical tasks must be designed and enabled by a clear understanding of customer-driven requirements that are fulfilled with rigor and discipline.

- **Scorecards vs. Checklists:** An integrated set of scorecards will help measure technical risk to improve decision making. Checklists must give way to more discriminating scorecards at the technical tool, task, and deliverable level. To properly make decisions, to "be in the know" about the fundamentals of risk, and to pay for great performance, you need summary data that goes beyond gut feelings, best guesses, and anecdotal opinion that gets mistaken for data.

Do a self-exam for hubris. If you have it, redouble your efforts to apply what you just read in this text. Just because lots of peers believe something to be true does not make it so. Your MBA, if you have one, is foundational and certainly not infallible. You're on a never-ending journey of learning and trying new patterns of structuring tasks and managing the resulting risks. Get in the game and study how your teams complete work. You will initially find they are *not* completing their work—they're just getting as much done as they can before they are forced to move on. This is the most common form of hubris we encounter. A lot of work is not getting done fully and completely in your organization. This is a different kind of lean problem. You are too lean. You get this way by doing too many projects with too few people and resources, which is a traditional lean problem.

I have never seen a technical team set out to be late or do an incomplete set of tasks that result in poor quality that they will inevitably have to clean up later. In fact, they despise getting into these situations. They prefer to do things right the first time. I sincerely hope this text has in some way stimulated you to take action to prevent problems on behalf of your teams. Your technical teams are looking to you for leadership, courage, vision, and, most of all, a realistic view of just how complex and difficult their jobs really become when they are given too many projects and not enough time to do any of them right.

We hope you have a better understanding of how Six Sigma and Lean affect the flow of work within and across technical processes in an enterprise. It will take a while for Six Sigma for Growth to have the same impact on the portfolio renewal and R&TD processes as it has on the commercialization and production community (typically associated with the well-known technical form of DFSS). Your personal attention, leadership, and willingness to take the hard fork in the road will make the difference as you seek the goal of sustainable growth.

GLOSSARY

A

affinity diagram
A tool used to gather and group ideas; usually depicted as a "tree" diagram.

ANOM
(Analysis of the mean) An analytical process that quantifies the mean response for each individual control factor level. ANOM can be performed on data that is in regular engineering units or data that has been transformed into some form of signal-to-noise ratio or other data transform. Main effects and interaction plots are created from ANOM data.

ANOVA
(Analysis of the variance) An analytical process that decomposes the contribution each individual control factor has on the overall experimental response. The ANOVA process also can account for the contribution of interactive effects between control factors and experimental error in the response if enough degrees of freedom are established in the experimental array. The value of epsilon squared (percentage of contribution to overall CFR variation) is calculated using data from ANOVA.

array

An arithmetically derived matrix or table of rows and columns that is used to impose an order for efficient experimentation. The rows contain the individual experiments. The columns contain the experimental factors and their individual levels or set points.

ASQ

American Society for Quality.

B

benchmarking

The process of comparative analysis between two or more concepts, components, subassemblies, subsystems, products, or processes. The goal of benchmarking is to qualitatively and quantitatively identify a superior subject within the competing choices. Often the benchmark is used as a standard to meet or surpass. Benchmarks are used in building Houses of Quality, concept generation, and the Pugh concept-selection process.

best practice

A preferred and repeatable action or set of actions completed to fulfill a specific requirement or set of requirements during the phases within a product-development process.

beta (β)

The Greek letter β is used to represent the slope of a best-fit line. It indicates the linear relationship between the signal factor(s) (critical adjustment parameters) and the measured critical functional response in a dynamic robustness-optimization experiment.

Black Belt

A job title or role indicating that the person has been certified as having mastered the Six Sigma DMAIC (Define-Measure-Analyze-Improve-Control) content and demonstrated expertise in leading one or more projects. The title usually designates the team leader of a Six Sigma project, often a coach of Green Belts.

blocking

A technique used in classical DOE to remove the effects of unwanted, assignable-cause noise or variability from the experimental response so that only the effects from the control factors are present in the response data. Blocking is a data-purification process used to help ensure the integrity of the experimental data used in constructing a statistically significant math model.

C

capability growth index (CGI)
The calculated percentage between 0 percent and 100 percent that a group of system, subsystem, or subassembly CFRs have attained in getting their Cp indexes to equal a value of 2 (indicating how well their critical functional responses have attained Six Sigma performance during product development). The CPI for critical functions is a metric often found on an executive gate review scorecard.

capability index
Cp and Cpk indexes that calculate the ratio of the voice of the customer versus the voice of the product or process. Cp is a measure of capability based on short-term or small samples of data—usually what is available during Product development. Cpk is a measure of long-term or large samples of data that include not only variation about the mean, but also the shifting of the mean itself—usually available during steady-state production.

checklist
A simple list of action items, steps, or elements needed to complete a task. Each item is checked off as it is completed.

classical design of experiments (DOE)
Experimental methods employed to construct math models relating a dependent variable (the measured critical functional response) to the set points of any number of independent variables (the experimental control factors). DOE is used sequentially to build knowledge of fundamental functional relationships (ideal/transfer functions) between various factors and a response variable.

commercialization
A business process that harnesses the resources of a company in the endeavor of conceiving, developing, designing, optimizing, certifying design and process capability, producing, selling, distributing, and servicing a product.

compensation
The use of feedforward or feedback-control mechanisms to intervene when certain noise effects are present in a product or process. Compensation is done only when insensitivity to noise cannot be attained through robustness optimization.

component
A single part in a subassembly, subsystem, or system. An example is a stamped metal part before it has anything assembled to it.

component requirements document

The document that contains all the requirements for a given component. They are often converted into a quality plan given to the production supply chain to set the targets and constrain the variation allowed in the incoming components.

control charts

A toolset used to monitor and control a process for variation over time, which varies with the type of data it monitors.

control factor

The factors or parameters (CFP or CTF specification) in a design or process that the engineer can control and specify to define the optimum combination of set points for satisfying the voice of the customer.

critical adjustment parameter (CAP)

A specific type of CFP that controls the mean of a CFR. These are identified using sequential DOE and engineering analysis. They are the input parameters for response surface methods for the optimization of mean performance optimization when robust design is completed. They enable Cpk to be set equal to Cp, thus enabling entitlement to be approached, if not attained.

critical functional parameter (CFP)

An input variable (usually an engineered additivity grouping) at the sub-assembly or subsystem level that controls the mean or variation of a CFR.

critical functional response (CFR)

A measured scalar or vector (complete, fundamental, continuous engineering variable) output variable that is critical to fulfilling a critical (highly important) customer requirement. Some refer to these critical customer requirements as CTQs. This metric is often found on an executive gate review scorecard.

critical parameter management (CPM)

The process that develops critical requirements and measures critical functional responses to design, optimize, and certify the capability of a product and its supporting network of manufacturing and service processes.

critical path

The sequence of tasks in a project that takes the greatest amount of time for completion.

critical-to-function specification (CTF)

A dimension, surface, or bulk characteristic (typically a scalar) that is critical to a component's contribution to a subassembly-, subsystem-, or system-level CFR.

criticality

A measurable requirement or functional response that is highly important to a customer. All requirements are important, but only a few are truly critical.

cross-functional team

A group of people representing multiple functional disciplines and possessing a wide variety of technical and experiential background and skills working together. Particularly applicable in the product commercialization process. (*See* multidisciplined teams.)

CTQ

Critical to quality, as defined by the customer.

D

dashboard

A summary and reporting tool of data and information about a process and/or product performance. Usually viewed as more complex than scorecards, and depicts the critical parameters necessary to run the business.

deliverable

Tangible, measurable output completed as an outcome of a task or series of tasks.

design capability (Cp$_d$)

The Cp index for a design's critical functional response in ratio to its upper and lower specification limits (VOC-based tolerance limits).

design of experiments (DOE)

A process for generating data that uses a mathematically derived matrix to methodically gather and evaluate the effect of numerous parameters on a response variable. When properly used, designed experiments efficiently produce useful data for model-building or engineering-optimization activities.

deterioration noise factor

A source of variability that results in some form of physical deterioration or degradation of a product or process. This is also referred to as an inner noise because it refers to variation inside the controllable factor levels.

DFSS

A Six Sigma concept used by the engineering technical community to design and develop a product. The acronym represents Design for Six Sigma.

DMADV

A five-step Six Sigma method used primarily to redesign a broken process, as well as to solve problems and/or improve processes or products of defects. The acronym stands for Define-Measure-Analyze-Design-Validate.

DMAIC

A five-step Six Sigma method used to solve problems and/or improve processes or products of defects. The acronym stands for Define-Measure-Analyze-Improve-Control.

DMEDI

A five-step method combining classic and Lean Six Sigma concepts to redesign a broken process, as well as to solve problems and/or improve processes or products of defects. The acronym stands for Define-Measure-Explore-Develop-Implement.

DPMO

Defects per million opportunities measurement.

E

economic coefficient

The economic coefficient is used in the quality loss function. It represents the proportionality constant in the loss function of the average dollars lost (A_0) due to a customer reaction to off-target performance and the square of the deviation from the target response (Δ_0^2). This is typically, but not exclusively, calculated when approximately 50 percent of the customers are motivated to take some course of economic action due to poor performance (but not necessarily outright functional failure). This is often referred to as the LD 50 point in the literature.

ECV

Expected commercial value; a financial metric often found on an executive gate review scorecard.

energy flow map

A representation of an engineering system that shows the paths of energy divided into productive and nonproductive work. This is analogous to a free body diagram, from an energy perspective. These maps account for the law of conservation of energy and are used in preparation of math modeling and design of experiments.

energy transformation

The physical process a design or product system uses to convert some form of input energy into various other forms of energy that ultimately produce a measurable response. The measurable response itself could be a form of energy or the consequence of energy transformations that have taken place within the design.

engineering metrics
A scalar or vector that is usually called a CTF specification, CFP, CAP, or CFR. They are greatly preferred over quality metrics (yield, defects, and so on) in DFSS.

engineering process
A set of disciplined, planned, and interrelated activities that engineers use to conceive, develop, design, optimize, and certify the capability of a new product or process design.

environmental noise factors
Sources of variability due to effects that are external to the design or product; also referred to as outer noise. They can also be sources of variability that one neighboring subsystem imposes on another neighboring subsystem or component. Examples include vibration, heat, contamination, misuse, and overloading.

experiment
An evaluation or series of evaluations that explore, define, quantify, and build data that can be used to model or predict functional performance in a component, subassembly, subsystem, or product. Experiments can be used to build fundamental knowledge for scientific research or to design and optimize product or process performance in the engineering context of a specific commercialization process.

experimental efficiency
This is a process-related activity that is facilitated by intelligent application of engineering knowledge and the proper use of designed experimental techniques. Examples include the use of fractional factorial arrays, control factors that are engineered for additivity, and compounded noise factors.

experimental error
The variability present in experimental data that is caused by meter error and drift, human inconsistency in taking data, random variability taking place in numerous noise factors not included in the noise array, and control factors that have not been included in the inner array. In the Taguchi approach, variability in the data due to interactive effects is often, but not always, included as experimental error.

experimental factors
Independent parameters that are studied in an orthogonal array experiment. Robust design classifies experimental factors as either control factors or noise factors.

experimental space

The combination of the entire control factor, noise factor, and signal factor (CAP) levels that produce the range of measured response values in an experiment.

F

F-ratio

The ratio formed in the ANOVA process by dividing the mean square of each experimental factor effect by the MS of the error variance. This is the ratio of variation occurring *between* each of the experimental factors in comparison to the variation occurring *within all the experimental factors* being evaluated in the entire experiment. It is a form of signal-to-noise ratio in a statistical sense. The noise in this case is random experimental error—not variability due to the assignable-cause noise factors in the Taguchi noise array.

feedback control system

A method of compensating for the variability in a process or product by sampling output response and sending a feedback signal that changes a critical adjustment parameter to put the mean of the response back on its intended target.

FMEA

Failure modes and effects analysis. A risk-analysis technique that identifies and ranks the potential failure modes of a design or process and then prioritizes improvement actions.

full factorial design

Two- and three-level orthogonal arrays that include every possible combination of experimental factors. Full factorial experimental designs use degrees of freedom to account for all the main effects and all interactions between factors included in the experimental array. Basically, all of the interactions beyond two-way interactions are likely to be of negligible consequence, so little need exists to use large arrays to rigorously evaluate such rare and unlikely three-way interactions and above.

fundamental

The property of a critical functional response that expresses the basic or elemental physical activity that is ultimately responsible for delivering customer satisfaction. A response is fundamental if it does not mix mechanisms and is uninfluenced by factors outside the component, subassembly, subsystem, and system design or production process being optimized.

G

Gantt chart
A horizontal bar chart used for project planning and control that lists the necessary project activities as row headings against horizontal lines showing the dates and duration of each activity.

gate
A short period of time during a process when the team reviews and reacts to the results against requirements from the previous phase and proactively plans for the smooth execution of the next phase.

gate review
Meeting with the project team and sponsors to inspect completed deliverables. The review focuses on the results from specific tools and best practices and manages the associated risks and problems. It also makes sure the team has everything it needs to apply the tools and best practices for the next phase with discipline and rigor. A gate review's time should be 20 percent reactive and 80 percent proactive.

goalpost mentality
A philosophy about quality that accepts anything within the tolerance band (USL-LSL) as equally good and anything that falls outside of the tolerance band as equally bad. See soccer, hockey, lacrosse, and football rulebooks.

goal statement
Identifies the critical parameters (including time frame) for a targeted improvement. (Use SMART technique to ensure completeness.)

GOSPA
Goals, objectives, strategies, plans, and actions planning methodology.

grand total sum of squares
The value obtained when squaring the response of each experimental run from a matrix experiment and then adding the squared terms.

Green Belt
A job title or role indicating that the person has been certified as having demonstrated an understanding of the basic Six Sigma DMAIC (Define-Measure-Analyze-Improve-Control) concepts. This role might support a Black Belt on a Six Sigma project or, in some companies, work on a small-scale project directly related to the job.

H

histogram
A graphical display of the frequency distribution of a set of data. Histograms display the shape, dispersion, and central tendency of the distribution of a data set.

House of Quality
An input/output relationship matrix used in the process of quality function deployment.

hypothesis testing
A statistical evaluation that checks the validity of a statement to a specified degree of certainty. These tests are done using well-known and quantified statistical distributions.

I

IDEA
A four-step Six Sigma method used by strategic marketing to define, develop, manage, and refresh a portfolio of offerings (products and services). The acronym represents Identify-Define-Evaluate-Activate.

ideal/transfer function
Fundamental functional relationships between various engineering control factors and a measured critical functional response variable. The math model of $Y = f(x)$ represents the customer-focused response that would be measured if there were no noise or only random noise acting on the design or process.

inbound marketing
Marketing activities that are focused on providing deliverables for internal consumption, as opposed to deliverables intended for the marketplace.

independent effect
The nature of an experimental factor's effect on the measured response when it is acting independently of any other experimental factor. When all control factors are producing independent effects, the design is said to be exhibiting an additive response.

inference
Drawing some form of conclusion about a measurable functional response based on representative or sample experimental data. Sample size, uncertainty, and the laws of probability play a major role in making inferences.

inner array
An orthogonal matrix that is used for the control factors in a designed experiment and is crossed with some form of outer noise array during robust design.

inspection
The process of examining a component, subassembly, subsystem, or product for off-target performance, variability, and defects during either product development or manufacturing. The focus is typically on whether the item under inspection is within the allowable tolerances. As with all processes, inspection itself is subject to variability, and out-of-spec parts or functions might pass inspection inadvertently.

interaction
The dependence of one experimental factor on the level set point of another experimental factor for its contribution to the measured response. Two types of interaction take place: synergistic (mild to moderate and useful in its effect) and antisynergistic (strong and disruptive in its effect).

interaction graph
A plot of the interactive relationship between two experimental factors as they affect a measured response. The ordinate (vertical axis) represents the response being measured, and the abscissa (horizontal axis) represents one of the two factors being evaluated. The average response value for the various combinations of the two experimental factors is plotted. The points representing the second factor's low level are connected by a line. Similarly, the points representing the second factor's next-higher level are connected by a line.

IRR
Internal rate of return (%IRR), a financial metric often found on an executive gate review scorecard.

K

KJ analysis
Stands for "Jiro Kawakita," a Japanese anthropologist who treated attributes (or language) data in a similar manner as variables of data by grouping and prioritizing it. A KJ diagram (similar to an affinity diagram) focuses on the unique and different output, linking the critical customer priorities to the project team's understanding and consensus.

L

lagging indicators

An indicator that follows the occurrence of something; hence, such indicators are used to determine the performance of an occurrence or an event. By tracking lagging indicators, one reacts to the results and determines, for example, the high and low temperature, precipitation, and humidity of a given day.

leading indicators

An indicator that precedes the occurrence of something; hence, such indicators are used to signal the upcoming occurrence of an event. By tracking leading indicators, one can prepare or anticipate the subsequent event and be proactive. For example, barometric pressure and doplar radar of a surrounding region are indicators of ensuing weather.

Lean Six Sigma

Modified Six Sigma approach to emphasize improving speed of a process by "leaning" it of its non-value-add steps. Typically used in a manufacturing environment, its common metrics include zero wait time, zero inventory, line balancing, batch sizes cut to improve flow through, and reduced overall process time.

level

The set point where a control factor, signal factor (CAP), or noise factor is placed during a designed experiment.

level average analysis

See ANOM (analysis of means).

lifecycle cost

The costs associated with making, supporting, and servicing a product or process over its intended life.

linearity

The relationship between a dependent variable (the response) and an independent variable (such as the signal or control factor) that is graphically expressed as a straight line. Linearity is typically a topic within the dynamic cases of the robustness process and in linear regression analysis.

linear combination

This term has a general mathematical definition and a specific mathematical definition associated with the dynamic robustness case. In general, a linear combination is the simple summation of terms. In the dynamic case, it is the specific summation of the product of the signal level and its corresponding response ($M_i y i_{i,j}$).

linear graph
A graphical aid used to assign experimental factors to specific columns when evaluating or avoiding specific interactions.

LMAD
A four-step Six Sigma method used by marketing to manage the ongoing operations of a portfolio of launched offerings (products and services) across the value chain. The acronym represents Launch-Manage-Adapt-Discontinue.

loss to society
The economic loss that society incurs when a product's functional performance deviates from its targeted value. The loss is often due to economic action the consumer takes when reacting to poor product performance, but it can also be due to the effects that spread through society when products fail to perform as expected. For example, a new car breaks down in a busy intersection due to a transmission defect and 14 people are 15 minutes late for work (cascading loss to many points in society).

lower specification limit
The lowest functional performance set point that a design or component can attain before functional performance is considered unacceptable.

M

main effect
The contribution an experimental factor makes to the measured response independent of experimental error and interactive effects. The sum of the half effects for a factor is equal to the main effect.

manufacturing process capability (Cp_m)
The ratio of the manufacturing tolerances to the measured performance of the manufacturing process.

matrix
An array of experimental set points that is derived mathematically. The matrix consists of rows (containing experimental runs) and columns (containing experimental factors).

matrix experiment
A series of evaluations that are conducted under the constraints of a matrix.

mean
The average value of a sample of data that is typically gathered in a matrix experiment.

mean square deviation (MSD)
A mathematical calculation that quantifies the average variation a response has with respect to a target value.

mean square error
A mathematical calculation that quantifies the variance within a set of data.

meter
A measurement device usually connected to some sort of transducer. The meter supplies a numerical value to quantify functional performance.

measured response
The quality characteristic that is a direct measure of functional performance.

measurement error
The variability in a data set that is due to poorly calibrated meters and transducers, human error in reading and recording data, and normal, random effects that exist in any measurement system used to quantify data.

Monte Carlo simulation
A computer-simulation technique that uses sampling from a random number sequence to simulate characteristics or events or outcomes with multiple possible values.

MSA
Measurement system analysis tool to understand the level of reproducibility and repeatability.

MTBF
Mean time between failure; measurement of the lapsed time from one failure to the next.

multidisciplined team
A group of people working together who possess a wide variety of technical and experiential background and skills. Particularly applicable in the product commercialization process. (*See* Cross-functional teams.)

N

noise
Any source of variability. Typically, noise is either external to the product (such as environmental effects) or a function of unit-to-unit variability due to manufacturing, or it can be associated with the effects of deterioration. In this context, noise is an assignable, nonrandom cause of variation.

noise directionality
A distinct upward or downward trend in the measured response, depending on the level at which the noises are set. Noise factor set points can be compounded, depending on the directional effect on the response.

noise experiment
An experiment designed to evaluate the strength and directionality of noise factors on a product or process response.

noise factor
Any factor that promotes variability in a product or process.

normal distribution
The symmetric distribution of data about an average point. The normal distribution takes on the form of a bell-shaped curve. It is a graphic illustration of how randomly selected data points from a product or process response will mostly fall close to the average response, with fewer and fewer data points falling farther and farther away from the mean. The normal distribution can also be expressed as a mathematical function and is often called a Gaussian distribution.

NPV
Net present value; a financial metric often found on an executive gate review scorecard.

NUD
Acronym representing new, unique, and difficult.

O

off-line quality control
The processes included in preproduction commercialization activities. The processes of concept design, parameter design, and tolerance design make up the elements of off-line quality control. It is often viewed as the area where quality is designed into the product or process.

online quality control
The processes included in the production phase of commercialization. The processes of statistical process control (loss function based and traditional), inspection, and evolutionary operation (EVOP) are examples of online quality control.

one factor at a time experiment
An experimental technique that examines one factor at a time determines the best operational set point, locks in on that factor level, and then moves on to repeat the process for the remaining factors. This technique is widely

practiced in scientific circles but lacks the circumspection and discipline provided by full and fractional factorial designed experimentation. Sometimes one-factor-at-a-time experiments are used to build knowledge prior to the design of a formal factorial experiment.

operating income
Calculated as gross profit minus operating expenses. A financial metric often found on an executive gate review scorecard.

operational marketing
Pertains to marketing's activities in support of launching and managing an offering (product and/or service) or set of offerings across the value chain.

optimize
Finding and setting control factor levels at the point where their mean, standard deviation or S/N ratios are at the desired or maximum value. Optimized performance means the control factors are set so that the design is least sensitive to the effects of noise and the mean is adjusted to be right on the desired target.

orthogonal
The property of an array or matrix that gives it balance and the capability of producing data that allow for the independent quantification of independent or interactive factor effects.

orthogonal array
A balanced matrix that is used to lay out an experimental plan for the purpose of designing functional performance quality into a product or process early in the commercialization process.

outbound marketing
Marketing activities that are focused on providing deliverables for the customers, as opposed to deliverables intended for internal consumption.

outer array
The orthogonal array used in dynamic robust design that contains the noise factors and signal factors. Each treatment combination of the control factors specified in the inner array is repeated using each of the treatment combinations specified by the outer array.

P

parameter
A factor used in the design, optimization, and certification of capability processes. Experimental parameters are CFRs, CFPs, CTF specs. and noise factors.

parameter design
The process employed to optimize the levels of control factors against the effect of noise factors. Signal factors (dynamic cases) or tuning factors (NTB cases) are used in the two-step optimization process to adjust the performance onto a specific target during parameter (robust) design.

parameter-optimization experiment
This is the main experiment in parameter design that is used to find the optimum level for the control factors. Usually, this experiment is done using some form of dynamic crossed array design.

PERT chart
Diagram that displays the dependency relationships between tasks.

phase
A period of time that is designed to conduct work to produce specific results that meet the requirements for a given project, wherein specific tools and best practices are used.

phase-gate product-development process
A series of time periods that are rationally divided into phases for the development of new products and processes. Gates are checkpoints at the end of each phase to review progress, assess risks, and plan for efficiency in future phase performance.

population parameter or statistic
A statistic such as the mean or standard deviation that is calculated with all the possible values that make up the entire population of data in an experiment. Samples are just a portion of a population.

probability
The likelihood or chance that an event or response will occur out of some number (n) of possible opportunities.

problem statement or opportunity for improvement
A clear, concise definition of what is wrong with a current process or product/services offering. This should be aligned with company strategies and/or annual plan. (Use SMART technique to ensure completeness.)

process
A set sequence of steps to make something or do something.

process capability analysis
Quantifies the capability of a process to produce output that meets customer requirements. Various capability metrics include DPMO (defects per million opportunities); Cp, Cpk (potential process capability, short term); Pp, Ppk (process capability, long term); and rolled throughput yield (RTY).

process map
A type of flowchart depicting the steps in a process, identifying its inputs outputs, and often assigning responsibility for each step and the key measures.

product commercialization
The act of gathering customer needs; defining requirements; conceiving product concepts; selecting the best concept; and designing, optimizing, and certifying the capability of the superior product for production, delivery, sales, and service.

product development
The continuum of tasks, from inbound marketing to technology development to certified technology being transferred into product design to the final step of the certified product design being transferred into production.

project cycle-time
The time that elapses from the beginning to the end of a project.

project management
The methods of planning, designing, managing, and completing projects. Project management designs and controls the micro timing of tasks and actions (underwritten by tools and best practices) within each phase of a product-development process.

Pugh process
A structured concept-selection process used by multidisciplinary teams to converge on superior concepts. The process uses a matrix consisting of criteria based on the voice of the customer and its relationship to specific candidate design concepts. The evaluations are made by comparing the new concepts to a benchmark called the datum. The process uses the classification metrics of "same as the datum," "better than the datum," and "worse than the datum." Several iterations are employed wherever increasing superiority is developed by combining the best features of highly ranked concepts until a superior concept emerges and becomes the new benchmark.

Q

QFD
Quality function deployment. A process for translating the voice of the customer into technical requirements at the product level. As part of the critical parameter-management process, QFD uses a series of matrixes called Houses of Quality to translate and link system requirements to subsystem requirements, which, in turn, are translated and linked to subassembly requirements, which are translated and linked to component requirements, which are translated and linked to manufacturing process requirements.

quality

The degree or grade of excellence. In a product-development context, it is a product with superior features that performs on target with low variability throughout its intended life. In an economic context, it is the absence or minimization of costs associated with the purchase and use of a product or process.

quality characteristic

A measured response that relates to a general or specific requirement that can be an attribute or a continuous variable. The quantifiable measure of performance that directly affects the customer's satisfaction. Often in DFSS, these have to be converted to an engineering scalar or vector.

quality engineering

Most often referred to as Taguchi's approach to off-line quality control (concept, parameter, and tolerance design) and online quality control.

quality function deployment (QFD)

A disciplined process for obtaining, translating, and deploying the voice of the customer into the various phases of technology development and the ensuing commercialization of products or processes during product design. *See* QFD.

quality loss cost

The costs associated with the loss to customers and society when a product or process performs off the targeted response.

quality loss function

The relationship between the dollars lost by a customer due to off-target product performance and the measured deviation of the product from its intended performance. Usually described by the quadratic loss function.

quality metrics

Defects, time to failure, yield, go/no go. *See* quality characteristic.

quality plan

The document used to communicate specifications to the production-supply chain. Often the component House of Quality is converted into a quality plan.

R

RACI matrix

A two-dimensional table that lists tasks or deliverables as the row headings, and roles (or people's names) as the column headings. The cell data contains the responsibility by task and by role (or person): R = Responsible; A = Accountable; C = Consulted; I = Informed.

random error
The nonsystematic variability that is present in experimental data due to random effects occurring outside the control factor main effects. The residual variation in a data set that was induced by unsuppressed noise factors and error due to human or meter error.

randomization
The technique employed to remove or suppress the effect of systematic or biased order (a form of assignable-cause variation) in running designed experiments. Randomizing is especially important when applying classical DOE in the construction of math models. It helps ensure that the data is as random as possible.

Real Win Worth (RWW)
Technique to analyze market potential, competitive position, and financial return. The composite value is often used as a summary metric on an executive gate review scorecard.

relational database
The set of requirements and fulfilling data that is developed and documented during critical parameter management. They link many-to-many relationships up and down throughout the hierarchy of the system being developed.

reliability
The measure of robustness over time. The length of time a product or process performs as intended.

repeat measurement
The taking of data points, where the multiple measured responses are taken without changing any of the experimental set points. Repeat measurements provide an estimate of measurement error only.

repeatability
Variation of repeated measurements of the *same item.*

replicate
The taking of data in which the design or process set points have all been changed since the previous readings were taken. Often a replicate is taken for the first experimental run and then again at the middle and end of an experiment (for a total of three replicates of the first experimental run). Replicate measurements provide an estimate of total experimental error.

reproducibility
1) The variation in the averages from repeated measurements made by *different people* on *the same item.* 2) The capability of a design to perform as targeted throughout the entire development, design, and production

phases of commercialization. Verification tests provide the data on reproducibility in light of the noise imposed on the design.

requirement
Criteria that must be fulfilled; something wanted or needed.

response
The measured value taken during an experimental run. Also called the quality characteristic. In DFSS, we prefer to focus on critical functional responses (CFRs).

risk mitigation
A planning process to identify, prevent, remove, or reduce risk if it occurs, and define actions to limit the severity or impact of a risk if it occurs.

ROI
Return on investment, calculated as the annual benefit divided by the investment amount. A financial metric often found on an executive gate review scorecard.

robust design
A process within the domain of quality engineering for making a product or process insensitive to the effects of variability without actually removing the sources of variability. Synonymous with *parameter design.*

S

sample
A select, usually random set of data points that are taken out of a greater population of data.

sample size
The measure of how many samples have been taken from a larger population. Sample size has a notable affect on the validity of making a statistical inference.

sample statistic
A statistic such as the mean or standard deviation that is calculated using a sample from the values that make up the entire population of data.

saturated experiment
The complete loading of an orthogonal array with experimental factors. The possibility exists that main effects will be confounded with potential interactions between the experimental factors within a saturated experiment.

scalar
A continuous engineering variable that is measured by its magnitude alone (no directional component exists for a scalar).

scaling factor
A critical adjustment parameter that is known to have a strong effect on the mean response and a weak effect on the standard deviation. The scaling factor often has the additional property of possessing a proportional relationship to both the standard deviation and the mean.

scaling penalty
The inflation of the standard deviation in a response as the mean is adjusted using a CAP.

scorecards
A set of critical summary data used to predict outcomes or evaluate performance of a process or product when making decisions and managing risk. Often called a dashboard.

screening experiment
Typically a small, limited experiment that is used to determine which factors are important to the response of a product or process. Screening experiments are used to build knowledge prior to the main modeling experiments in sequential DOE methodology.

sensitivity
The change in a CFR based upon unit changes in a CFP, a CAP, or a CTF spec. Also a measure of the magnitude (steepness) of the slope between the measured response and the signal factor (CAP) levels in a dynamic robustness experiment.

Sigma (σ)
The standard deviation (technically, a measure of the population standard deviation).

signal-to-noise ratio
A ratio or value formed by transforming the response data from a robust design experiment using a logarithm to help make the data more additive. Classically, signal-to-noise is an expression relating the useful part of the response to the nonuseful variation in the response.

signal factor
A critical adjustment parameter that is known to be capable of adjusting the average output response of the design in a linear manner. Signal factors are used in dynamic robustness cases as well as in response surface methods.

SIPOC
A summary tool to communicate the suppliers, inputs, process, outputs, and customers of a process.

Six Sigma (6s)

A disciplined approach to enterprise-wide quality improvement and variation reduction. Technically, it is the denominator of the capability (Cp) index.

slope

Quantifies the linear relationship between the measured response and the signal factor (CAP). *See* beta.

smaller the better (STB)

A static case in which the smaller the measured response is, the better the quality of the product or process is.

SMART problem and goal statement

An acronym specifying that a project problem and goal statement needs to be specific, measurable, achievable (but aggressive), relevant (to the project team and business), and time-bounded.

SPC

Statistical process control is a method for assessing data stability and "fit" as related to an expected normal probability distribution that is assumed to represent the shape of dispersion of the sample data.

specification

A specific quantifiable set point that typically has a nominal or target value and a tolerance of acceptable values associated with it. The values result when a team tries to use tools and best practices to fulfill a requirement. Requirements are hopefully completely fulfilled by a final design specification, but many times they are not.

sponsor

A role indicating that the person has ultimate accountability for a Six Sigma project, its direction, and its funding. This person is a senior manager whose job is highly dependent on the outcome of the project—often the functional process owner of the process on which the project is focused. The project team reports to the sponsor for this project. The sponsor conducts gate reviews of the project and serves as the project's liaison to the company's goals and missions.

standard deviation

A measure of the variability in a set of data. It is calculated by taking the square root of the variance. Standard deviations are not additive; the variances are.

static robustness case
One of the two major types of experimental robustness cases to study a product or process response as related to specific design parameters. The static case has no predesigned signal factor associated with the response. Thus, the response target is considered fixed or static. Control factors and noise factors are used to find local optimum set points for static robustness performance.

strategic marketing
Pertains to marketing's activities in support of the definition, development, management, and refresh of a portfolio of offerings (products or services) at an enterprise, business unit, or division level.

subassembly
Any two or more components that can be assembled into a functioning assembly.

subassembly requirements document
A document that contains the requirements, both critical and all others, for a subassembly.

subsystem
A group of individual components and subassemblies that perform a specific function within the total product system. Systems consist of two or more subsystems.

subsystem requirements document
A document that contains the requirements, both critical and all others, for a subsystem.

sum of squares
A calculation technique used in the ANOVA process to help quantify the effects of the experimental factors and the mean square error (if replicates have been taken).

sum of squares due to the mean
The calculation of the sum of squares to quantify the overall mean effect due to the experimental factors being examined.

supply chain
The network of suppliers that provides raw materials, components, subassemblies, subsystems, software, or complete systems to your company.

surrogate noise factor
Time, position, and location are not actual noise factors, but they stand in nicely as surrogate "sources of noise" in experiments that do not have clearly defined physical noises. These are typically used in process robustness-optimization cases.

synergistic interaction

A mild to moderate form of interactivity between control factors. Synergistic interactions display monotonic but nonparallel relationships between two control factors when their levels are changed. They are typically not disruptive to robust design experiments.

system

An integrated group of subsystems, subassemblies, and components that make up a functioning unit that harmonizes the mass, energy and information flows, and transformations of the elements to provide an overall product output that fulfills the customer-based requirements.

system integration

The construction and evaluation of the system from its subsystems, subassemblies, and components.

system requirements document

A document that contains the requirements, both critical and all others, for a system.

T

tactical marketing

Pertains to marketing's activities in support of the design and development of an offering (product or service) to make it ready for commercialization or launch.

Taguchi, Genichi

The originator of the well-known system of quality engineering. Dr. Taguchi is an engineer, former university professor, author, and global quality consultant.

target

The ideal point of performance that is known to provide the ultimate in customer satisfaction. Often called the nominal set point or the ideal performance specification.

task

A specific piece of work (or definable activity) required to be done as a duty.

technology development

The building of new or leveraged technology (R&D) in preparation for transfer of certified technology into product design programs.

technology transfer
The hand-off of certified (robust and tunable) technology and data-acquisition systems to the design organization.

tool
An instrument or device that aids the completion of a task.

total sum of squares
The part of the ANOVA process that calculates the sum of squares due to the combined experimental factor effects and the experimental error. The total sum of squares is decomposed into the sum of squares due to individual experimental factor effects and the sum of squares due to experimental error.

transfer function
Fundamental functional relationships between various engineering control factors and a measured critical functional response variable. The math model of $Y = f(x)$ represents the customer-focused response that would be measured if there were no noise or if only random noise were acting on the design or process. Sometimes they are called transfer functions because they help model how energy, mass, and logic and control signals are transferred across system, subsystem, and subassembly boundaries.

treatment combination
A single experimental run from an orthogonal array.

two-step optimization process
The process of first finding the optimum control factor set points to minimize sensitivity to noise and then adjusting the mean response onto the customer-focused target.

U

UAPL
A four-step Six Sigma method used by tactical marketing to ready an offering (products and services) for commercialization. The acronym represents Understand-Analyze-Plan-Launch. This process is designed to be conducted in parallel with the technical community developing a product or service.

UMC
Unit manufacturing cost. The cost associated with making a product or process.

unit-to-unit variability
Variation in a product or process due to noises in the production and assembly process.

upper specification limit

The largest functional performance set point that a design or component is allowed before functional performance is considered unacceptable.

V

variance

The mean squared deviation of the measured response values from their average value.

variation

Changes in parameter values due to systematic or random effects. Variation is the root cause of poor quality and the monetary losses associated with it.

vector

An engineering measure that has both magnitude and directionality associated with it. Vectors are highly valued metrics for critical parameter management.

verification

The process of validating the results from a model or a designed experiment.

value chain

For a launched offer, the Go-To-Market value chain describes those functions that add value (according to the customer or client) to an already launched offering (product or service)—for example, sales, value-added resellers, other distribution channels, consultants (delivering incremental software and services), service (break-fix, maintenance, parts), training, customer support call center, and marketing support materials and advertising.

voice of the business (VOB)

The internal requirements of the business in the words of senior or executive management. The VOB is used throughout the product commercialization process and must be balanced with customer requirements.

voice of the customer (VOC)

The wants and needs of customers, in their own words. The VOC is used throughout the product commercialization process to keep the requirements and the designs that fulfill them focused on the needs of the customer.

voice of the process

The measured functional output of a manufacturing process.

voice of the product
The measured functional output of a design element at any level within the engineered system.

W

WCBF
Worldwide Conventions and Business Forums.

work break-down structure (WBS)
The process of dividing a project into manageable tasks and sequencing them to ensure a logical flow between tasks.

workflow charts
Diagrams that depict the flow of work in a process.

Y

Y = f (x)
A mathematical equation read as "Y equals f of x", which means the result (output) measures (represented by Y) are a function of the process (input) measures (represented by x). This equation is used to understand how the inputs affect the outputs.

yield
The percentage of the total number of units produced that are acceptable. Percent good.

INDEX

A

ACTIVATE phase (IDEA process), 60, 64, 86-87
ADAPT phase (LMAD process), 298, 308-310
additivity and run designed experiments, Optimize phase
    CDOV roadmap, 233-234
    I²DOV, 141-142
affinity diagramming, 179
analysis
    analysis of means (ANOM), 142
    analysis of performance of concepts, 124-125
    analysis of variance (ANOVA), 138, 154
    Design phase (CDOV roadmap), 212
ANOM (analysis of means), 142
ANOVA (analysis of variance), 138, 154
applications, growth, 2-6

B

balanced resources, designed cycle-time, 39
benchmarking (technology), Invention/Innovation phase (I²DOV), 109
best practices
    Concept phase (CDOV roadmap), 193-194
    Design phase (CDOV roadmap), 222-223
    Develop phase (I²DOV), 128-129
    Invention/Innovation phase (I²DOV), 116
    Optimize phase
        CDOV roadmap, 240, 248-249
        I²DOV, 144
    Verify phase
        CDOV roadmap, 260-261, 270
        I²DOV, 156
book of requirements, 258
book of translated requirements, 258

C

CAP (critical adjustment parameter), 284
capability growth index (CGI), 259
capability of tools (tool scorecards), 25-26
capability studies (I²DOV)
    Develop phase, 126-128
    Verify phase, 156
CDOV process, 16, 168
    Concept phase
        deliverables, 172
        integration table, 194
        prerequisite information, 193
        readiness, 192
        requirements, 171
        tasks, 172-192
        tools and best practices, 193-194
    Design phase
        deliverables, 197-198
        integration table, 223
        readiness, 221
        requirements, 197
        tasks, 198-221
        tools and best practices, 222-223
    Fast Track Projects, 279
        Concept phase, 280-283, 290
        Design phase, 283-285, 290
        Optimize phase, 285-287, 290
        Verify phase, 287-291
    major elements, 171
    Optimize phase, 226
        deliverables, 226
        integration table, 240-242
        phase 3B, 243-249
        readiness, 239
        requirements, 226
        tasks, 227-239
        tools, methods, and best practices, 240
    Verify phase, 249
        deliverables, 252
        integration table, 261

phase 4B, 261, 266-270
readiness, 260
requirements, 252
tasks, 253-259
tools and best practices, 260-261
certification of adjustment factors,
    Optimize phase (I²DOV), 133
CGI (capability growth index), 259
characterization
    Design phase (CDOV roadmap), 212
    Verify phase (I²DOV), 156
charts
    Gantt, 41
    I-MR, 285
    PERT, 41
checklists, 23, 321
collaborative innovation, 320
commercialization, Fast Track Projects,
    276-277
    DMAIC competencies, 277-279
    macrosummary, 289-290
    technology development, 279-289
competence centers, 319
competitive product benchmarking,
    181-182
compounded noises, Optimize phase
    CDOV roadmap, 232-233
    I²DOV, 138-139
concept evaluation criteria, Develop
    phase (I²DOV), 123
Concept phase (CDOV roadmap)
    deliverables, 172
    integration table, 194
    prerequisite information, 193
    readiness, 192
    requirements, 171
    risk profiles/tool-task recommendations,
        280-283, 290
    tasks, 172-192
    tools and best practices, 193-194
conduct SPC and capability studies,
    126-128
construction, technology roadmaps, 106
context
    DMADV process, 8, 11
    DMAIC process, 7-8, 11
    DMEDI process, 9-11
contextual inquiry, 177
controls, technical support processes,
    303-304
core competence centers, 319
critical adjustment parameter (CAP), 284
critical functional parameter nominal set
    points, 143
critical parameter management, 99, 184
critical tasks, designed cycle-time, 39
criticality, 218
current technologies (technology
    roadmaps), 106

cycle-time (projects), 36
    project management, 36-40
        designed cycle-time, 40-42
        failure modes, 46-49
        *integrated project cycle-time
            model, 49*
        Monte Carlo simulations, 42-44
    realistic design of, 321

D
data
    analysis
        Optimize phase (I²DOV), 142
        technical support processes, 303-304
    evaluation, Verify phase (I²DOV), 153
    integrity, tool scorecards, 24
    sets, LMAD process, 301-302
data-acquisition systems, Verify phase
    (I²DOV), 152
DEFINE phase (IDEA process), 60-63,
    72-78
deliverable-based processes, 19
deliverables
    CDOV roadmap
        Concept phase, 172
        Design phase, 197-198
        Optimize phase, 226
        Verify phase, 252
    I²DOV
        Develop phase, 120
        Invention/Innovation phase, 103
        Optimize phase, 132
        Verify phase, 147
    IDEA process
        ACTIVATE phase, 86
        DEFINE phase, 77
        EVALUATE phase, 81-82
        IDENTIFY phase, 65-66
    LMAD process
        ADAPT phase, 309
        DISCONTINUE phase, 311-312
        LAUNCH phase, 305
        MANAGE phase, 307
design failure modes and effects analysis
    (DFMEA), 218
design for additivity, 125
design for patentability, 107
Design phase (CDOV roadmap)
    deliverables, 197-198
    integration table, 223
    readiness, 221
    requirements, 197
    risk profiles/tool-task recommendations,
        283-285, 290
    tasks, 198-221
    tools and best practices, 222-223
designed cycle-time, 38-40
    failure modes, 46-49
    key steps, 40-42
    Monte Carlo simulations, 42-44
    realistic design of, 321

Develop phase (I²DOV)
  best practices, 128-129
  deliverables, 120
  gate 2 readiness, 128
  integration table, 129
  methods, 128-129
  modeling, 121
  personnel, 122
  requirements, 120
  tasks, 121-128
  tools, 128-129
development phase project plans
      (I²DOV), 115
DFMEA (design failure modes and
      effects analysis), 218
DFSS, 2
DISCONTINUE phase (LMAD process),
      298, 310-312
DMADV process, 8, 11
DMAIC process, 2, 7-11, 277-279,
      295-298
DMEDI process, 9-11
documentation
  CFPs and CFR relationships, 155
  Invention/Innovation phase
      (I²DOV),106, 112
  failure modes, 46-49

E
easy data sets (LMAD process), 301-302
ECO functions, 291
engineering control factors, Optimize
      phase
  CDOV roadmap, 233
  I²DOV, 140-141
engineering support processes, 294
  DMAIC process, 295-298
  LMAD process, 295-297
      ADAPT phase, 298, 308-310
      defining technical support processes,
          302-303
      DISCONTINUE phase, 298, 310-312
      hard versus easy data sets, 301-302
      LAUNCH phase, 298, 305-306
      MANAGE phase, 298, 307
      milestone scorecards, 313
      models, data analysis, and controls,
          303-304
      phase-milestone flow, 299-300
      risk management, 303
environmental stress screening
      (ESS), 247
ESS (environmental stress screen-
      ing), 247
EVALUATE phase (IDEA process),
      60-63, 81-85
executive-level gate review scorecards,
      29-31
exposure assessment, 117
external forms, 62

F
failure modes, designed cycle-time, 46-49
family plans, defining, 108
Fast Track Projects, 19, 276-277
  DMAIC competencies, 277-279
  macrosummary, 289-290
  technology development, 279-289
FMEA project format, 47-49
functional-level gate review scorecards,
      29-31
functions, 110
fundamental technical variables, 318
future technologies (technology
      roadmaps), 106
future trends, 318-321

G
Gantt charts, 41
gates, 11
  defined, 60
  designed cycle-time, 38
  I²DOV readiness
      Invention/Innovation phase, 115-116
      Develop phase, 128
      Optimize phase, 143
      Verify phase, 155-156
  passage, 13
  review scorecards, 29-31
  reviews, 11
  three-tiered system of colors, 13, 28-29
generation, Invention/Innovation phase
      (I²DOV)
  development phase project plan, 115
  risk assessments and summary profiles,
      114-115
  technology concepts that fulfill
      functions, 113-114
  technology system requirements
      documents, 110
geometric models, 216
growth
  future trends, 318-321
  processes, 15-19
  Six Sigma applications, 2-6

H
HALT (highly accelerated life-testing),
      219, 247
hard data sets, LMAD process, 301-302
HAST (highly accelerated stress testing),
      219, 247
historical technologies (technology
      roadmaps), 106
House of Quality
  Concept phase (CDOV roadmap),
      179-180
  Invention/Innovation phase (I²DOV),
      108-109

I–J

I-MR charts, 285

I^2DOV, 15
  phase-gate approach to R&TD, 100
    Develop phase, 120-129
    Invention/Innovation, 103-117
    Optimize phase, 132-146
    project management steps, 102
    Verify phase, 147-158

IDEA (Identify, Define, Evaluate, and
    Activate), 15, 53-55, 60-63
  ACTIVATE phase, 64, 86-87
  DEFINE phase, 63, 72-78
  EVALUATE phase, 63, 81-85
  IDENTIFY phase, 60-68, 72
  Task scorecard, 90

IDENTIFY phase (IDEA process),
    60-68, 72

in-bound marketing, 318

inbound growth enablement, 4

inbound work, portfolio renewal process,
    52-53
  phases of work (IDEA), 53-90
  process discipline, 55-59

initial reliability model, Concept phase
    (CDOV roadmap), 191-192

Innovation phase (I^2DOV)
  best practices, 116
  deliverables, 103
  gate 1 readiness, 115-116
  integration table, 117
  methods, 116
  patent analysis and exposure
      assessment, 117
  requirements, 103
  tasks, 103-115
  tools, 116

integrated project cycle-time model, 49

integration table
  Concept phase (CDOV roadmap), 194
  Design phase
    CDOV roadmap, 223
    I^2DOV, 129
  Optimize phase
    CDOV roadmap, 240-242, 249
    I^2DOV, 144-146
  Verify phase
    CDOV roadmap, 261, 270
    I^2DOV, 156, 158

integrity of data (tool scorecards), 24

interactions, 212

internal forms, 62

Invention phase (I^2DOV)
  best practices, 116
  deliverables, 103
  gate 1 readiness, 115-116
  integration table, 117
  methods, 116
  patent analysis and exposure
      assessment, 117

requirements, 103
tasks, 103-115
tools, 116

K–L

KANO model of customer satisfaction,
    177-178

KJ analysis, 178-179

latent needs (customers), 107

LAUNCH phase (LMAD process), 298,
    305-306

Lean Six Sigma, Fast Track Projects,
    276-277
  DMAIC competencies, 277-279
  macrosummary, 289-290
  technology development, 279-289

linking discoveries, 112

LMAD process, 16, 295-297
  ADAPT phase, 298, 308-310
  defining technical support processes,
      302-303
  DISCONTINUE phase, 298, 310-312
  hard versus easy data sets, 301-302
  LAUNCH phase, 298, 305-306
  MANAGE phase, 298, 307-308
  milestone scorecards, 313
  models, data analysis, and controls,
      303-304
  phase-milestone flow, 299-300
  risk management, 303

M

macrosummary, Fast Track Projects,
    289-290

MANAGE phase (LMAD process), 298,
    307-308

management
  project management, 36-38
    designed cycle-time, 38-44
    documentation of failure modes,
        46-49
    integrated project plan, 49
  risk management
    checklists, 23
    scorecards, 22-32

math models, 216

measurement systems, Design phase
    (CDOV roadmap), 212-214

methods
  CDOV roadmap
    concept phase 171-197
    Design phase, 222-223
    Optimize phase, 240, 248-249
    Verify phase, 270
  I^2DOV
    Develop phase, 128-129
    Invention/Innovation phase, 116

milestone scorecards, 313

modeling
  Design phase (CDOV roadmap), 212
  Develop phase (I^2DOV), 121

models
  integrated project cycle-time model, 49
  process integration, 17
  technical support processes, 303-304
Monte Carlo simulations, designed
      cycle-time, 42-44
multi-tasking, 319

N

noise diagrams, Optimize phase
  CDOV roadmap, 229-231
  I²DOV, 136-137
noise factor experiments, Optimize phase
  CDOV roadmap, 231-232
  I²DOV, 137-138
noises (post-launch), 297
nominal optimal performance, 139
nominal performance tests, Verify phase
      (I²DOV), 152
nonlinear effects experiments, 214
NUD functions, 281

O

operational post-launch engineering
      support processes, 294
  DMAIC process, 295, 298
  LMAD process, 295-297
    ADAPT phase, 298, 308-310
    defining technical support processes,
        302-303
    DISCONTINUE phase, 298, 310-312
    hard versus easy data sets, 301-302
    LAUNCH phase, 298, 305-306
    MANAGE phase, 298, 307
    milestone scorecards, 313
    models, data analysis, and controls,
        303-304
    phase-milestone flow, 299-300
    risk management, 303
Optimize phase
  CDOV roadmap, 226
    deliverables, 226
    integration table, 240-242
    phase 3B, 243-249
    readiness, 239
    requirements, 226
    risk profiles/tool-task
        recommendations, 285-290
    tasks, 227-239
    tools, methods, and best
        practices, 240
  I²DOV
    best practices, 144
    certification of adjustment
        factors, 133
    deliverables, 132
    gate 3 readiness, 143
    integration table, 144-146
    requirements, 132
    robustness optimization, 133
    tasks, 132-143
    tools, 144-146

outbound growth enablement, 4
outbound post-launch engineering, 3

P

P&TPR (product and technology
      portfolio renewal), IDEA, 53-90
  ACTIVATE phase, 64, 86-87
  DEFINE phase, 63-78
  EVALUATE phase, 63, 81-85
  IDENTIFY phase, 61-72
  Task scorecard, 90
participation strategy, 181
passage, gates, 13
patent analysis, 117
performance improvement, 154
PERT charts, 41, 170
phase-gate processes, 11
  checklists, 23
  gates, 11
    passage, 13
    reviews, 11
      three-tiered system of colors, 13,
          28-29
  scorecards, 22-32
phase-milestone flow (LMAD process),
      299-300
phases
  defined, 60
  LMAD process
    ADAPT phase, 298, 308-310
    DISCONTINUE phase, 298, 310-312
    LAUNCH phase, 298, 305-306
    MANAGE phase, 298, 307
  work phases (IDEA), 53-90
    ACTIVATE phase, 64, 86-87
    DEFINE phase, 63, 72-78
    EVALUATE phase, 63, 81-85
    IDENTIFY phase, 61-72
    Task scorecard, 90
physical models, 216
platform development, 110
platform noise maps, 136-137
platforming, 320
PLS (product line strategies), 108
portfolio management (LMAD
      process), 16
post-launch engineering support
      processes, 294
  DMAIC process, 295-298
  LMAD process, 295
    ADAPT phase, 298, 308-310
    defining technical support processes,
        302-303
    DISCONTINUE phase, 298, 310-312
    hard versus easy data sets, 301-302
    LAUNCH phase, 298, 305-306
    MANAGE phase, 298, 307
    milestone scorecards, 313
    models, data analysis, and controls,
        303-304
    phase-milestone flow, 299-300
    risk management, 303

predictive additive model, 234-235
prerequisite information, Concept phase
(CDOV roadmap), 193
probability of occurrence term
(DFMECA), 218
problem-solving steps
DMADV process, 8, 11
DMAIC process, 7-8, 11
DMEDI process, 9-11
processes
CDOV, Fast Track Projects, 279-289
deliverable-based, 19
discipline, 55-59
DMADV, 8, 11
DMAIC, 2, 7-8, 11
DMEDI, 9-11
growth, 15-19
integration model, 17
maps, designed cycle-time, 40
operational post-launch engineering
support, 294
DMAIC process, 295-298
LMAD process, 295-313
phase-gate processes, 11
checklists, 23
gates, 11-13, 28-29
scorecards, 22-32
project management, 36-38
designed cycle-time, 38-44
documentation of failure modes,
46-49
integrated project plan, 49
R&TD, 96-100
critical parameter management, 99
I²DOV roadmap, 100-158
strategic product and technology
portfolio renewal, 52-53
phases of work (IDEA), 53-90
process discipline, 55-59
tactical product commercialization,
164-165
CDOV roadmap, 168-270
preparation, 165-168
product and technology portfolio
renewal. *See* P&TPR
product commercialization
R&TD, I²DOV roadmap, 100-158
tactical product commercialization
process, 164-165
CDOV roadmap, 168-270
preparation, 165-168
product line strategies (PLS), 108
product test fixtures, 150
product-commercialization (CDOV
process), 16
project management, 36-38
cycle-time, 36
designed cycle-time, 40-42
failure modes, 46-49
integrated project cycle-time
model, 49

key steps, 40-42
Monte Carlo simulations, 42-44
documentation of failure modes, 46-49
Fast Track Projects, 19, 276-277
DMAIC competencies, 277-279
macrosummary, 289-290
technology development, 279-289
FMEA, 47-49
I²DOV approach to R&TD, 102
Develop phase, 120-129
Invention/Innovation, 103-117
Optimize phase, 132-146
Verify phase, 147-158
integrated project plans, 49
realistic design, 321

Q–R
QFD process, 179
quality of tools (tool scorecards), 24

R&TD (research and technology
development process), 2, 96-100
critical parameter management, 99
I²DOV roadmap, 100
Develop phase, 120-129
Invention/Innovation, 103-117
Optimize phase, 132-146
project management steps, 102
Verify phase, 147-158
RACI matrix, 41
readiness
CDOV roadmap
Concept phase, 192
Design phase, 221
Optimize phase, 239, 248
Verify phase, 260, 269
I²DOV
Develop phase, 128
Invention/Innovation phase, 115-116
Optimize phase, 143
Verify phase, 155-156
refining discoveries, Invention/Innovation
phase (I²DOV), 112
reliability models (CDOV roadmap)
Concept phase, 191-192
Design phase, 217-220
requirements
CDOV roadmap
Concept phase, 171
Design phase, 197
Optimize phase, 226
Verify phase, 252
I²DOV
Develop phase, 120
Invention/Innovation phase, 103
Optimize phase, 132
Verify phase, 147
IDEA process
ACTIVATE phase, 86
DEFINE phase, 72-73
EVALUATE phase, 81
IDENTIFY phase, 62-65

LMAD process
  ADAPT phase, 309
  DISCONTINUE phase, 311
  LAUNCH phase, 305
  MANAGE phase, 307
research and technology development
  process. *See* R&TD
resolution path, 13
response surface experiments, 214,
  236-237
reviews, gates, 11, 29-31
risk assessment, 114-115
risk management
  checklists, 23
  scorecards, 22-32
  technical support processes, 303
risk priority number (RPN), 49
risk profiles (CDOV process)
  Concept phase, 280-283, 290
  Design phase, 283-285, 290
  Optimize phase, 285-287, 290
  Verify phase, 287-291
robust experiments, Optimize phase
  CDOV roadmap, 232-233
  I^2DOV, 133, 138-139
rolled throughput yield (RTY), 259
RPN (risk priority number), 49
RTY (rolled throughput yield), 259

S
scorecards, 22-23, 32, 321
  gate review scorecards, 29-31
  LMAD process, 313
  task scorecards, 26-29, 90
  tool scorecards, 23-26
screening experiments, 214
*Setting the PACE in Product
  Development,* 17
Six Sigma for Marketing Processes, 175
solving problems
  DMADV process, 8, 11
  DMAIC process, 7-8, 11
  DMEDI process, 9-11
SPC studies, Develop phase (I^2DOV),
  126-128
specific product requirements, 64
SRD (system requirements
  document), 183
SS CFRs, 151
SS nominal performance tests, 152
Steps, designed cycle-time, 40-42
strategic product and technology
  portfolio renewal process, 52-53
  phases of work (IDEA), 53-90
    ACTIVATE phase, 64, 86-87
    DEFINE phase, 63-78
    EVALUATE phase, 63, 81-85
    IDENTIFY phase, 61-72
    Task scorecard, 90
  process discipline, 55-59

strategic research and technology
  development process. *See* R&TD
strategic technical process environment
  (IDEA process), 60-90
  ACTIVATE phase, 64, 86-87
  DEFINE phase, 63, 72-78
  EVALUATE phase, 63, 81-85
  IDENTIFY phase, 61-72
  Task scorecard, 90
stress tests, Verify phase (I^2DOV),
  153-154
sublevel benchmarking, 201-203
sublevel requirements documents,
  203-204
subsystem House of Quality, 199-201
subsystem noise diagrams, 136-137
summary profiles, 114-115
superior system concepts, 189-190
superior technology concepts, 124
support and conduct invention, 111-112
system concepts, 188-189
system noise maps, 229-231
system of colors, gates, 13, 28-29
system requirements document
  (SRD), 183
system-level functions, 183-185
Systems Architecting, 187

T
tactical product commercialization
  process, 164
  CDOV roadmap, 168-171
    Concept phase, 171-194
    Design phase, 197-223
    major elements, 171
    Optimize phase, 226-249
    Verify phase, 249-270
  preparation, 165-168
tactical product design-engineering
  process, 16
task scorecards, 26-27, 90
tasks
  CDOV roadmap
    Concept phase, 172-192
    Design phase, 198-221
    Optimize phase, 227-248
    Verify phase, 253-269
  Gantt charts, 41
  I^2DOV
    Develop phase, 121-128
    Invention/Innovation phase, 103-115
    Optimize phase, 132-143
    Verify phase, 147-155
  IDEA process
    ACTIVATE phase, 87
    DEFINE phase, 77
    EVALUATE phase, 82
    IDENTIFY phase, 66
  LMAD process
    ADAPT phase, 310

DISCONTINUE phase, 312
LAUNCH phase, 306
MANAGE phase, 308
performance of
checklists, 23
scorecards, 22-32
resolution path, 13
TDFSS (Technology Development for Six
Sigma), 2
technical task performance
checklists, 23
scorecards, 22-32
technical variables, 318
technical workflow design, 294
DMAIC process, 295-298
LMAD process, 295
ADAPT phase, 298, 308-310
defining technical support processes,
302-303
DISCONTINUE phase, 298, 310-312
hard versus easy data sets, 301-302
LAUNCH phase, 298, 305-306
MANAGE phase, 298, 307
milestone scorecards, 313
models, data analysis, and controls,
303-304
phase-milestone flow, 299-300
risk management, 303
technology development
Fast Track Projects, 279-289
I²DOV process, 15
Technology Development for Six Sigma
(TDFSS), 2
technology system requirements
documents, 110
technology trends, 106
three-tiered system of colors, gates, 13,
28-29
tool scorecards, 23
capability of tools, 25-26
integrity of data, 24
quality of tools, 24
tool-task recommendations (CDOV
process)
Concept phase, 280-283, 290
Design phase, 283-285, 290
Optimize phase, 285-287, 290
Verify phase, 287-291
tool-task-deliverable sets, 39
tools
CDOV roadmap
Concept phase, 193-194
Design phase, 222-223
Optimize phase, 240, 248-249
Verify phase, 260-261, 270
I²DOV
Develop phase, 128-129
Invention/Innovation phase, 116
Optimize phase, 144-146
Verify phase, 156

IDEA process
ACTIVATE phase, 87
DEFINE phase, 78
EVALUATE phase, 83
IDENTIFY phase, 67
Total Quality Development, 177
trained resources, designed cycle-
time, 39
training, investment in, 318
translating the voice of the customer,
107-108
trends for the future, 318-321

U–V
value-adding tasks, phases, 11
verification experiments (Optimize
phase)
CDOV roadmap, 235-238
I²DOV, 143
verification of reduced sensitivities, 155
Verify phase
CDOV roadmap, 249
deliverables, 252
integration table, 261
phase 4B, 261, 266-270
readiness, 260
requirements, 252
risk profiles/tool-task
recommendations, 287-291
tasks, 253-259
tools and best practices, 260-261
I²DOV
best practices, 156
capability studies and
characterization, 156
deliverables, 147
gate 4 readiness, 155-156
integration table, 156-158
requirements, 147
tasks, 147-155
tools, 156
VOC (voice-of-the-customer) data,
107-108, 173-177

W–Z
Winning at New Products, 17
work break-down structures, designed
cycle-time, 40
work phases (IDEA), 53-90
ACTIVATE phase, 64, 86-87
DEFINE phase, 63, 72-78
EVALUATE phase, 63, 81-85
IDENTIFY phase, 61-68, 72
Task scorecard, 90
workflow charts, 40

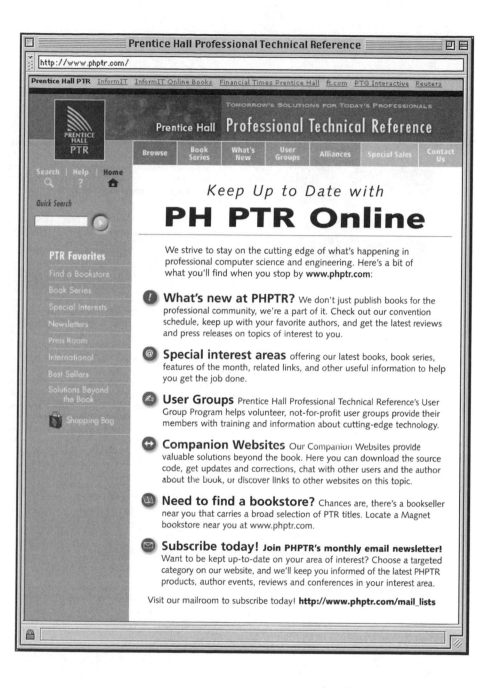

For everyone seeking to maximize quality in the development and design of any technology product or service!

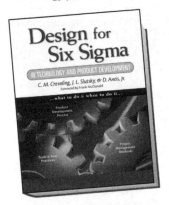

Design For Six Sigma in Technology and Product Development

BY CLYDE CREVELING, JEFF SLUTSKY, AND DAVE ANTIS

"The authors of this book have worked as designers and consultants leading the transition from build, test, and fix to disciplined, fact-based designs that delight customers and stakeholders alike. I am not aware of any other book that discusses Design for Six Sigma in such a comprehensive and practical way as this one. This is the right book for leaders and designers who want to change from hoping for the best to expecting the best."
—Steve Schaus, VP of Operational Excellence, Sequa Corporation

Design for Six Sigma is the first book to show companies how to tightly link design for Six Sigma (DFSS) to the phases and gates of a well-structured product development process, and carefully manage it through a rigorous project management discipline.

© 2003, Cloth, 800 pages, 0130092231

Tolerance Design

BY CLYDE M. CREVELING

Tolerance Design has become a highly valued tool in the DFSS methods portfolio. The procedures outlined in this seminal text on the analytical and empirical development of tolerances enable worst case analysis, Root Sum of Squares analysis, Monte Carlo simulation, sensitivity analysis, capability studies, tolerance trade-off studies, and cost vs. quality balancing. Use of Designed Experiments in the optimization of tolerances is a main area of focus that this book brings to life through a number of practical examples. *Tolerance Design* helps answer the pivotal question, "Just what level of sigma performance do my systems, functional assemblies, parts, and manufacturing processes really need to meet critical-to-customer quality and performance requirements?" DFSS practitioners around the globe use this helpful text to do the right things at the right time during their tolerance design and validation activities.

© 1997, Cloth, 448 pages , 0201634732

Six Sigma for Marketing Processes

BY CLYDE M. CREVELING AND LINDA HAMBLETON

This book provides an overview of the way marketing professionals can utilize the value offered by Six Sigma tools, methods, and best practices. It provides unique methods for employing Six Sigma to enhance the three marketing processes for enabling a business to attain growth: strategic, tactical, and operational.

© 2006, Cloth, 304 pages, 013199008X

Learn how to use Taguchi Methods and other robust design techniques that focus on engineering processes to optimize technology and products for better performance under the imperfect conditions of the real world.

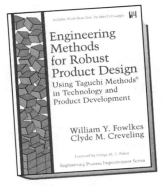

Engineering Methods for Robust Product Design

BY WILLIAM Y. FOWLKES AND CLYDE M. CREVELING

Quality in products and product-related processes is now, more than ever, a critical requirement for success in manufacturing. This book offers simple, yet effective, guidelines on how to practice robust design in the context of a total quality development effort. With practical techniques, and real-life examples, this hands-on book teaches practicing engineers and students how to use Taguchi methods along with other robust design techniques such as Six Sigma to improve processes and designs.

© 1995, Cloth, 432 pages, 0201633671

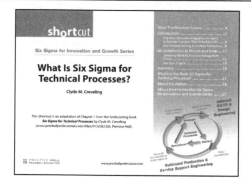

What Is Six Sigma for Technical Processes? (Digital Short Cut)

BY CLYDE M. CREVELING

This PDF document discusses how technical leaders and management professionals can implement Six Sigma for the processes they oversee. This form of Six Sigma focuses on four process arenas that enable a business to attain a state of sustainable growth: Strategic Portfolio Renewal Process, Strategic R&TD Process, Tactical Design Engineering Process, and Operational Production and Service Support Engineering Process.

© 2007, Electronic, 27 pages, 0131574221